THE CORE MODEL

THE CORE MODEL

A Collaborative Paradigm for the Pharmaceutical Industry and Global Health Care

IBIS SÁNCHEZ-SERRANO
President & Founder The Core Model Corporation, S.A. (C.M.C.) Panamá City, Panamá

Academic Press is an imprint of Elsevier
125 London Wall, London EC2Y 5AS, United Kingdom
525 B Street, Suite 1650, San Diego, CA 92101, United States
50 Hampshire Street, 5th Floor, Cambridge, MA 02139, United States
The Boulevard, Langford Lane, Kidlington, Oxford OX5 1GB, United Kingdom

Notices

Knowledge and best practice in this field are constantly changing. As new research and experience broaden our understanding, changes in research methods, professional practices, or medical treatment may become necessary.

Practitioners and researchers must always rely on their own experience and knowledge in evaluating and using any information, methods, compounds, or experiments described herein. In using such information or methods they should be mindful of their own safety and the safety of others, including parties for whom they have a professional responsibility.

To the fullest extent of the law, neither the Publisher nor the authors, contributors, or editors, assume any liability for any injury and/or damage to persons or property as a matter of products liability, negligence or otherwise, or from any use or operation of any methods, products, instructions, or ideas contained in the material herein.

Library of Congress Cataloging-in-Publication Data

A catalog record for this book is available from the Library of Congress

British Library Cataloguing-in-Publication Data

A catalogue record for this book is available from the British Library

ISBN: 978-0-12-814293-6

For information on all Academic Press publications visit our website at https://www.elsevier.com/books-and-journals

Publisher: Andre G. Wolff
Acquisition Editor: Erin Hill-Parks
Editorial Project Manager: Megan Ashdown
Production Project Manager: James Selvam
Cover Designer: Mark Rogers

Typeset by TNQ Technologies

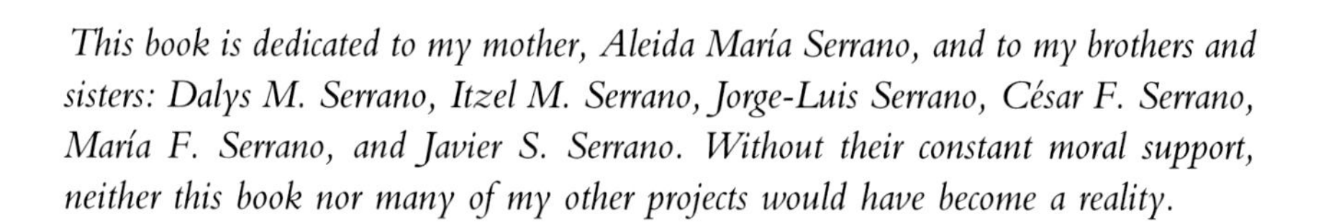

This book is dedicated to my mother, Aleida María Serrano, and to my brothers and sisters: Dalys M. Serrano, Itzel M. Serrano, Jorge-Luis Serrano, César F. Serrano, María F. Serrano, and Javier S. Serrano. Without their constant moral support, neither this book nor many of my other projects would have become a reality.

I would also like to dedicate this book to my good friend Daniel Morand, of Switzerland, who made this book as well as The World's Health Care Crisis *a possibility through his generous financial support and indefatigable encouragement.*

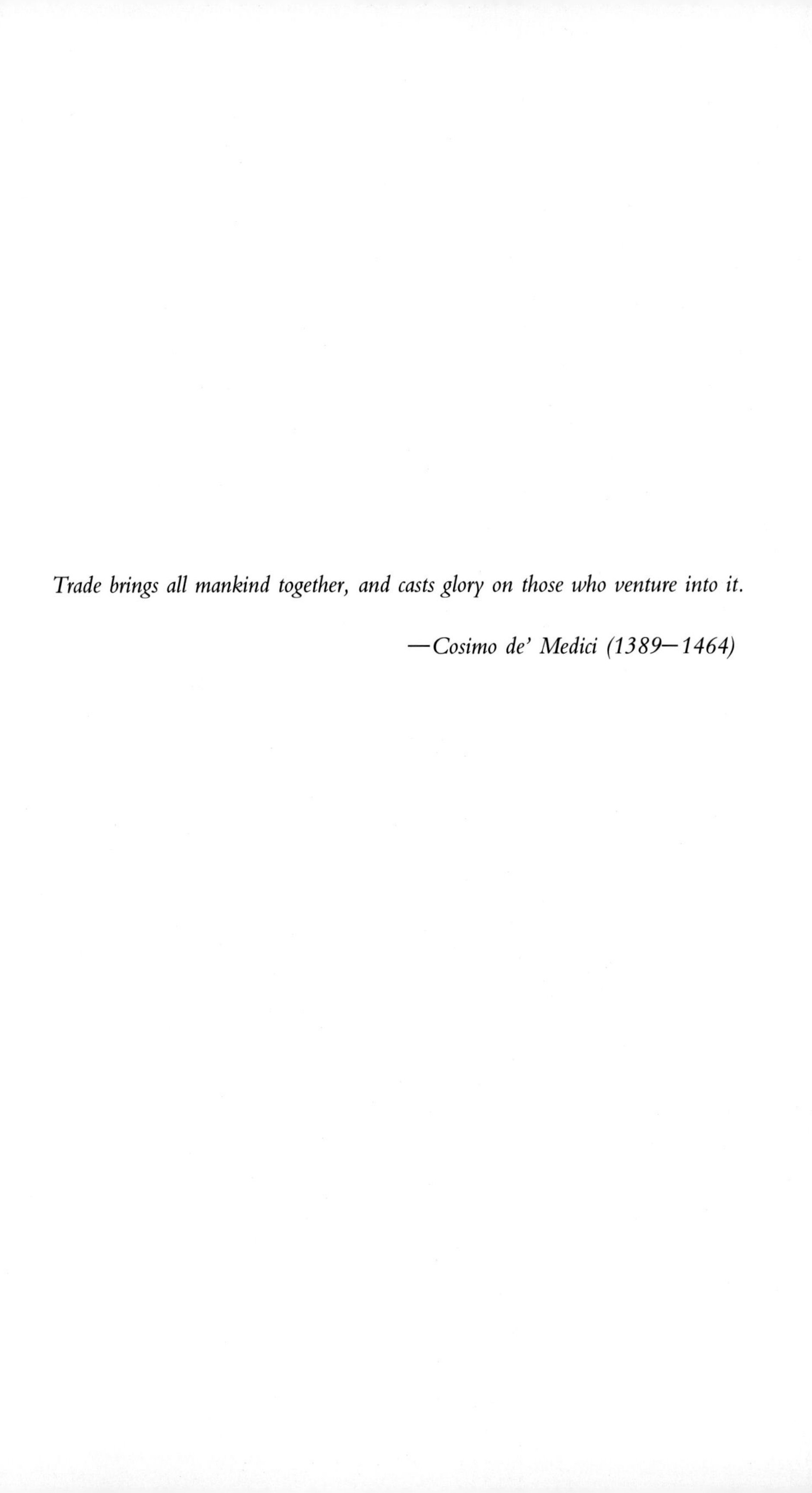

Trade brings all mankind together, and casts glory on those who venture into it.

—Cosimo de' Medici (1389–1464)

Contents

Professional profile: Ibis Sánchez-Serrano

Ibis Sánchez-Serrano is a world-leading expert on global health care, pharmaceutical innovation, and translational science policy. He is the author of two groundbreaking books *The World's Health Care Crisis: From the Laboratory Bench to the Patient's Bedside* (Elsevier, 2011) and *The Core Model: A Collaborative Paradigm for the Pharmaceutical Industry and Global Health Care* (Academic Press/Elsevier, June 2019). *The World's Health Care Crisis* is the first comprehensive analysis ever of the current global health care situation and the role of the pharmaceutical industry within it. This work has been praised as "excellent" by *Choice* magazine, nominated to the world-class Spanish "Prince of Asturias" Award, recipient of a prestigious Fulbright "Recognition" Award, and subject to significant high-profile media coverage worldwide. *The Core Model* is a pragmatic organizational model to optimize the process of drug discovery and development, with important implications for global health care.

Mr. Sánchez-Serrano has built a reputation among important international government bodies as an expert on the world's health care crisis and the relevant solutions. For example, he has been invited by the U.S. National Institutes of Health (NIH; National Center for Advancing Translational Sciences/ Therapeutics for Rare or Neglected Diseases

[TRND] Program) to present and discuss his views on the US health care system, the significant challenges associated with the high cost of pharmaceutical R&D, and the high price of innovative drugs. Recently, he has been invited by Novartis Institutes for Biomedical Research (NIBR), Cambridge, Massachusetts, to present his research on Gender-Specific/ Precision Medicine and its importance in drug discovery, development, and global health care. Given his unique expertise and success in these fields, he is exploring entrepreneurial opportunities in drug discovery and development.

Sánchez-Serrano is the discoverer of "The Core Model," an economic and organizational paradigm for R&D, originally described in his 2006 paper, "Success in Translational Research: Lessons from the Development of Bortezomib", published in *Nature Reviews Drug Discovery*. Considered a major breakthrough, this concept has been developed further in his contributing central chapter to the book *Collaborative Innovation in Drug Discovery: Strategies for Public and Private Partnerships* (edited by Rathnam Chaguturu, Wiley, 2014), written with other thought-leaders in the field in academia and industry from around the globe, as well as in his most recent book, which will be published in 2019 by Elsevier. Furthermore, Ibis has broadened his range of research by investigating the field of pharmaceutical development in Gender-Specific Medicine. As a result, he was requested to write a fundamental book chapter entitled "Gender-Specific Medicine in Drug Discovery and Development" for the third edition of the classic medical textbook *Principles of Gender-Specific Medicine* (edited by Marianne Legato; Elsevier, 2017; Professional and Scholarly Excellence [PROSE] Award for the best book in Clinical Medicine, 2018).

Sánchez-Serrano has been interviewed extensively by the international media, including Cable-Satellite Public Affairs Network (C-SPAN), Cable News Network (CNN), UNIVISION, EFE Agency, Associated Press (AP), Deutsche Presse Agentur (DPA), El Tiempo and Teleantioquia (Colombia), and the major media outlets of his native Panamá.

An internationally sought-after speaker, Ibis has been an Invited Lecturer at the New York Public Library (broadcasted nationwide by C-SPAN radio and TV); Invited Participant at the fourth World Justice Project Forum (The Hague, Netherlands); and Invited Speaker at the fifth International Conference on Drug Discovery and Therapy (Dubai, United Arab Emirates [UAE]). In addition, Ibis has been Speaker at the First Official Conference of the International Chemical Biology Society, at the Novartis Research Institutes in Cambridge, Massachusetts, USA; and "Speaker of the Month" at the Fulbright Alumni Association monthly lecture series (Smithsonian Tropical Research Institute, Panamá). Ibis has been Keynote Speaker at the United States Agency for International Development (USAID) Closing Ceremony After 50 Years of Operations in Panamá. He has also been Keynote Speaker at the Biotechnology Executives (BioExec) Institute eighth Annual Retreat (Miami, Florida); Guest Expert on Global Infectious Diseases at INDICASAT AIP Panamá; and Guest Speaker at Technology, Entertainment, and Design Talks (TEDx) Panamá City, Panamá. He has been the Chairman at the First Enabling Future Pharma Conference track "Pre-Competitive Collaborative Innovation Strategies" in Chicago, Illinois, and Special Guest Speaker at "Farmacosmética, 2014," in Colombia, among many other speaking engagements. For four consecutive years (2014—17), Ibis has acted as a Judge for the Massachusetts Institute of Technology (MIT) Technology Review "Innovators Under 35" Award (Central America, Argentina, Uruguay, Mexico, Colombia, and Perú) in the Life Sciences/ Medicine sector. Recently, Ibis was a Speaker and the Moderator of the Forum "Evolution of an Academic Discovery: First Person Account of the Development of Velcade," organized and sponsored by Massachusetts Biotechnology Council (MassBio; Cambridge, Massachusetts) based on Ibis's 2006 *Nature Reviews Drug Discovery* article on the development of bortezomib. The forum comprised some of the key discoverers and developers of this breakthrough drug. Other recent engagements include Keynote Speaker at "The Inaugural 'Advancing Drug Development Forum - Making the Impossible Possible'" (Boston, MA), and Distinguished Guest Speaker at the Luis "Chicho" Fábrega (Panamá) on "The Importance of Gender-Specific Medicine in Pharmacology and Clinical Practice" and "Obesity, Genetics, Life-Style, and the Global Public Health Crisis." In 2019, Ibis was a Special Guest Speaker at NIBR, (Cambridge, Massachusetts) on the subject of Gender-Specific Medicine and Pharmaceutical Drugs Development.

Ibis has been invited to meet former U.S. President Bill Clinton and former U.S. Assistant Secretary of State for Public Affairs, Mike Hammer, as well as a significant number of high-profile scientific, political, business, and diplomatic figures from around the world. In September 2015, he was invited to meet the Minister of Health of Panamá, the Adviser of the Panamanian President for Health Care, and the Panamanian Health Care Cabinet and to present to them a Magisterial Talk on Global Health Care and Pharmaceutical Policy. In December 2015, he was a Keynote Speaker at the First Global Men's Health Summit (held in Panamá City, Panamá) and a "Witness of Honor" of the Summit's Resolution. Ibis received a

"Guest of Honor" Recognition Scroll by the City Council of Panamá City, along other figures such as Nobel Laureate Aaron Ciechanover, for his work on global health care.

In 2017, the Panamanian Minister of Health and his team asked Ibis his opinion on two bill proposals on the regulation of medical devices, and health care research in Panamá, respectively. He has also conceived and proposed the creation of the first National Genetics Research Institute in Panamá.

Ibis Sánchez-Serrano holds a B.S. in Genetics and Art History from Iowa State University, and a Masters Degree in International Business Relations and Technology Management from the Tufts Fletcher School of Law and Diplomacy in collaboration with Harvard University and MIT's Sloan School of Management (where he completed his thesis). Ibis has conducted graduate studies and research into oncology, molecular genetics, and biotechnology at the University of Pavia (Italy), the Pasteur Institute (France), Iowa State University, and Boston University. He has co-authored several papers on molecular genetics in international peer-reviewed scientific journals and is a former Fulbright Scholar (Caribbean and Latin America Scholarship Program [CLASP] II/Thomas Jefferson Fellow), a former Organization of American States (OAS) Fellow, and is a current member of the Editorial Board of the "Gender and the Genome" journal.

For his work, in March 2017, Ibis was selected by Uruguay's GEN Center for the Arts & Sciences as one of "The Top Twelve Latin American Scientists Changing the World." As a result, he was invited to discuss his research in a documentary produced by GEN/Santa Rita Films (Brazil)/ NBCUniversal (USA) on the current state of science in Latin America, soon to be broadcasted across the entire continent.

Ibis speaks four languages, enjoys photography and nature, and is an art, classical music, and literature connoisseur and an art collector. He is based in Boston, Massachusetts, USA, and Panamá City, Panamá, where he is very involved in community development and educational activities. In March 2018, Ibis Sánchez-Serrano was awarded by the Municipality of his native city, Santiago de Veraguas, Panamá, the honorable distinction "Meritorious Son of the City of Santiago de Veraguas," one of the most-prestigious recognitions given to a Panamanian Citizen. In November 2018, Sánchez-Serrano received, from the Municipality of Santiago de Veraguas, Panamá, another recognition award for his work in the Life Sciences. He has also received other awards for his work on global health care.

Ibis is an expert on the life and work of Italian Baroque composer Antonio Vivaldi and on the Renaissance polymath Leonardo da Vinci. At present, he is working on a book on Leonardo da Vinci.

Selected authorship

- *The Core Model: A Paradigm for the Pharmaceutical Industry and for Global Health Care* (Academic Press/Elsevier, 2019). [Book]
- "Disruptive Approaches to Accelerate Drug Discovery and the Development" (Co-authored with Tom Pfeifer and Rathnam Chaguturu). *Drug Discovery World.* Spring 2018.
- "Gender-Specific Medicine's Imminent Coming of Age". *Gender and the Genome* Journal (Invited Essay. December 2017).
- "The Remarkable Story of the Development of Velcade and the Model that Explains Its Success". *MassBio Blog*, July 19, 2017. https://www.massbio.org/news/blog/the-remarkable-story-of-the-development-of-velcade-and-the-model-that-explains-its-success-134330
- "Gender-Specific Medicine in Drug Discovery and Development" (Book Chapter in *Principles of Gender-Specific Medicine*, Third Edition, Edited by Marianne Legato; Academic Press, 2017. PROSE Award, 2018).
- "The Core Model: Drug Discovery and Development via Effective Translational Science and Public-Private Collaboration" (Book Chapter in *Collaborative Innovation in Drug Discovery*, First Edition, Edited by Rathnam Chaguturu; Wiley, 2014).
- "In the Global Health Care Crisis, Pharma is Crucial". *ElsevierConnect*, 3 September 2013; https://www.elsevier.com/connect/in-the-global-health-care-crisis-pharma-is-crucial
- *The World's Health Care Crisis: From the Laboratory Bench to the Patient's Bedside* (Elsevier, 2011). [Book; translated into Spanish]
- "Success in Translational Research: Lessons from the Development of Bortezomib" (*Nature Reviews Drug Discovery*, February 2006).

Selected media

- March 28, 2018, 8.00 a.m. NEXTV Noticias Panamá: Interviewed by Fernando Correa on the importance of Gender-Specific Medicine in Clinical Practice. https://www.youtube.com/watch?v=IJuoFHM5iY8

- July 26, 2017, 8:00—10:00 a.m., Cambridge, MA: Forum "Evolution of an Academic Discovery: First Person Account of the Development of Velcade" at MassBio. https://www.massbio.org/events/event-archive/evolution-of-an-academic-discovery-first-person-account-of-the-development-of-velcade-2101
- Video of the Forum "Evolution of an Academic Discovery: First Person Account of the Development of Velcade" http://www.ustream.tv/recorded/106250511
- March 23, 2014, 8:00/11:00 p.m., EST, Washington, D.C.: Ibis Sánchez-Serrano is interviewed by Brian Lamb in his high-profile interview program Q & A. Broadcasted worldwide by radio, TV, and Internet. http://www.c-span.org/video/?317867-1/qa-ibis-snchezserrano
- December 24—27 & 30, 2013, Manhattan, NY/Washington, D.C.: Ibis Sánchez-Serrano's book talk at the Mid-Manhattan Public Library (New York Public Library) on the global health care crisis is broadcasted by C-SPAN (Radio, TV, & Internet). http://www.c-span.org/video/?314786-1/pharmaceuticals-global-health-care-policy
- November 23, 2012, 6:00 p.m., Raleigh, NC.: Ibis Sánchez-Serrano is interviewed by Emmy Award winner Edwin Pitti of U.S.-based giant Hispanic network UNIVISION on his book "The World's Health Care Crisis". http://www.facebook.com/photo.php?v=277796785656752
- October 10, 2012, Panama City, Panamá: Invited Speaker. TEDx Panama City, 2012 on the "Core Model for Drug Discovery and Development". https://www.youtube.com/watch?v=ZQaq5_MJYjk&t=12s
- March 1, 2012 9:00 p.m., Miami, FL: Ibis Sánchez-Serrano was interviewed by Ismael Cala, in his Program CALA, CNN. 9:00 p.m. EST. http://www.youtube.com/watch?v=fEKks83kEH8

Contact information

Email: ibis.sanchezserrano@gmail.com

Preface

The purpose for writing this book is manifold. Primarily, I would like to bring to a wider audience the organizational paradigm/economic theory for pharmaceutical development that I discovered in 2004, while still a graduate student,[1] named the "Core Model." Initially, this model was published in 2006 in the prestigious scientific journal *Nature Reviews Drug Discovery*,[2] in which I explained how using it (unknowingly), Cambridge-based biotechnology company Myogenics/ProScript (later acquired by Millennium Pharmaceuticals) brought to the market the anticancer drug bortezomib (Velcade), for the treatment of multiple myeloma. This happened not only in record time, but also at a much lower cost than the industry's average.[3] For the first time, a model that could comprehend the academia–industry relationship, the private–public sectors interaction, basic-applied and translational research and the process of drug discovery and development was presented.

One of the major findings of the Core Model—and I would like to emphasize this—is that scientific collaboration in society does not occur in an "unstructured" manner. Actually, it takes place within a highly structured order, which I describe as the "Core," the "Bridge," and the "Periphery," through which knowledge is transferred, integrated, and finally translated into commercial products for the benefit of society. The way in which this is accomplished is via a mechanism that I call "trade of assets"—that is, exchange of personal connections, information, knowledge, experimental results, etc., among collaborating parties. In fact, "trade of assets" among collaborators is the driving force of innovation and, if done correctly, could save an incredible amount of time, labor, and economic resources when developing a new drug. The understanding of this model and its application (by the private and public sectors) could help us solve the global pharmaceutical industry's productivity problem and address important global health care and economic problems, such as having affordable access to high-quality innovative drugs. The Core Model, which deals with fostering innovation, could, in fact, be applied to other industries besides the life sciences.

Since its initial publication, I have been requested to republish the Core Model in several different ways: as an abridgement in Discovery Magazine[4] and in translated form into Spanish by the College of Pharmaceutical

Chemists in Colombia.[5] As a book chapter, the Core Model appeared in the book "Collaborative Innovation in Drug Discovery" (Wiley, 2014; edited by Rathnam Chaguturu, for which my article served as an inspiration, according to the editor).[6] It has been the subject of a Technology, Entertainment, and Design Talks (TEDx) talk[7]; and it has appeared in a recent article on disruptive approaches to drug discovery and development.[8] In addition, I have been invited to present this model at different important global venues, including the Fifth International Conference on Drug Discovery and Therapy (Dubai, 2013)[9]; and the First Official Conference of the International Chemical Biology Society (which took place at the Novartis Institute for Biomedical Research, in Cambridge, Massachusetts, 2014).[10] The model has been presented at the National Center for the Advancement of Translational Sciences (NCATS) of the US National Institutes of Health (NIH) (Bethesda, Maryland, 2014); and at Massachusetts Biotechnology Council (MassBio) (Cambridge, Massachusetts, 2017), together with the discoverers and developers of bortezomib[11]; among other places. All of this attests to the "popularity" of the model, even if within a small group of key opinion leaders. I must add, however, that over the years it has become clear to me that the Core Model is not only applicable to the life sciences and the pharmaceutical industry, but that it is actually a universal paradigm without which innovation and economic growth cannot be understood. Therefore, it deserves a longer study and exposition that justifies the writing of a book on it, while addressing an important economic problem: global health care.

In fact, in the last decade, I have seen how some companies, other than ProScript/Millennium (as illustrated in Chapter 7), are successfully applying the model; how the industry is naturally adopting the model; and how institutions such as the NIH are implementing it. The latter was illustrated by the 2014 decision of the NIH to collaborate with 10 Big Pharma companies to develop new drugs. As a result, the model has been enriched over the last decade and the time has come to present it fully developed to the global life sciences, health care, and economics communities.

In my book, *The World's Health Care Crisis: From the Laboratory Bench to the Patient's Bedside* (Elsevier, 2011), I defined the world's health care crisis as a fundamentally economic and then as a scientific crisis. I identified the pharmaceutical component as a crucial element in the crisis of the global health care systems. Within this scenario, two of the major hurdles in pharma are the high cost of R&D and the outdated pharmaceutical industry business model, based on financial interests—at the expense of societal well-being—and on secrecy, at the expense of knowledge generation, transfer, and integration, which are essential to progress in science. All of this has a tremendous and deleterious effect not only on the pharmaceutical industry itself, but also on global health care, as in the end the patient and the countries' economies are the ones that have to pay for the high bill. This, in fact, goes against the actual world economic goal of commercializing scientific research for the benefit of taxpayers. Therefore, to make important progress in our efforts to solve the world's health care crisis, we need to find better ways to develop pharmaceuticals and learn more and faster about the biology of human disease. This can only be accomplished by using new models that will bring optimization of all the resources available in society to that end, which is indeed what the Core Model does. Therefore, this book is in many ways a continuation of my previous Elsevier book *The World's Health Care Crisis*, while being an autonomous and completely independent and different book. Nowadays, when the pharmaceutical industry does not know where it is headed, when further reforms to the US health care system are imminent in the new "Trump Era," which intends to override the Affordable Care Act (ObamaCare), and when many of the health care[12] systems in the world are collapsing, it is the best timing for this work.

This book is organized in four different parts. The first part, Chapters 1 and 2, describe the remarkable story case of the development of bortezomib and the key people involved and introduce the Core Model. The second part, Chapters 3, 4, and 5, delves more deeply into each one of the individual components of the Core Model (Core, Bridge, Periphery), using the development of bortezomib as an example. The third part of the book, highlights the importance of the Core Model vis-à-vis intellectual property, the current state of the pharmaceutical industry, and global health care; and the fourth part explores the Core Model as novel economic theory that could be applied not only in the biopharmaceutical industry, but also in other industries.

By providing both general and very detailed views of innovation, collaboration, science, technology, the biopharmaceutical industry, and global health care, this book, written in plain English, will bring up-to-date information, story cases, and provocative insights to its readers. Readers will include entrepreneurs, biotech start-up companies, pharmaceutical companies, academic researchers, federal funding agencies, drug-making regulators, investors, philanthropic organizations, health care workers and practitioners, students, economists and patient advocacy groups. This book reveals what drug discovery and development and health care are really about. It is my hope that this work will be a major contribution to both science and economics.

Ibis Sánchez-Serrano
Boston, Massachusetts, United States
Santiago de Veraguas, Panamá

Endnotes

1. Sánchez-Serrano, I., 2004. Academia–Biotech Spin-Offs: What Is Inside the Black-Box? Translational Research and Academia–Industry Collaborations in the Process of Drug Development (Thesis). Tufts University and Massachusetts Inst. Technology.
2. Sánchez-Serrano, I., 2006. Success in translational research: lessons from the development of bortezomib. Nature Reviews Drug Discovery, 5 (2),107–114.
3. Year 2003.
4. Sánchez-Serrano, I., July 26, 2009. Translational research in the development of bortezomib: a core model. Discovery Medicine. http://www.discoverymedicine.com/Ibis-Sanchez-serrano/2009/07/26/translational-research-in-the-development-of-bortezomib-a-core-model/.
5. http://www.opinion.com.bo/opinion/articulos/2013/1019/noticias.php?id=109263.
6. Sánchez-Serrano, I., 2014. "The Core Model: Drug Discovery and Development via Effective Translational Science and Public–Private Collaboration" (Chapter 35) in Chaguturu, R. Collaborative Innovation in Drug Discovery. Wiley. https://onlinelibrary.wiley.com/doi/book/10.1002/9781118778166.
7. Sánchez-Serrano, I. 10 October 2012. El Modelo Core para el desarrollo de medicamentos. TEDx, Panama City. https://www.youtube.com/watch?v=ZQaq5_MJYjk (with English subtitles).
8. Sánchez-Serrano, I., Pfeifer, T., Chaguturu, R., 2018. Disruptive Approaches to Accelerate Drug Discovery and Development (Part 1). Drug Discovery World, Spring 2018. https://www.ddw-online.com/business/p322101-disruptive-approaches-to-accelerate-drug-discovery-and-development-(part-1).html.
9. International Conference on Drug Discovery and Therapy, 18–21 February 2013; Dubai, United Arab Emirates. http://www.eurekaselect.com/node/122425/5supthsup-international-conference-on-drug-discovery-and-therapy-2013.
10. First Official Conference of the International Chemical Biology Society, Novartis Institutes for Biomedical Research, Cambridge, MA, 4–5 October 2012. https://www.chemical-biology.org/page/MeetingArchive.
11. First Person Account: The Remarkable Story of the Development of Bortezomib, MassBio, Cambridge, MA, 2013; https://www.massbio.org/news/blog/the-remarkable-story-of-the-development-of-velcade-and-the-model-that-explains-its-success-134330.
12. Sánchez-Serrano, I., 2011. The World's Health Care Crisis: From the Laboratory Bench to the Patient's Bedside. Elsevier.

Acknowledgments

I would like to express my profound gratitude to Erin Hill–Parks (Acquisition Editor), Megan Ashdown (Editorial Project Manager), Kattie Washington (Acquisition Editor), and James Selvam (Production Project Manager & Team) at Elsevier for their interest in this project and for their patience.

From the very beginning during my graduate student days at Tufts University (in collaboration with Harvard and Massachusetts Institute of Technology (MIT), where I was a cross-registered student) my friends Daniel Morand and Nurettin Demirdöven believed in the Core Model and encouraged me to write the original *Nature Reviews Drug Discovery* article, thinking that it was something really novel and, perhaps, revolutionary. Later, Sam Williams, in London, and Rathnam Chaguturu, in New Jersey, became two other big enthusiasts of the model and, together with Daniel Morand, encouraged me to write this book. My brother César F. Serrano, who has a very sharp mind, also thought that I needed to write this book. To all of them, my heartfelt thanks.

I would also like to thank Peter Kirkpatrick, Editor-in-Chief at *Nature Reviews Drug Discovery* for deciding to publish the original paper; and my final thanks go to the Countway Library of Medicine of the Harvard Medical School, in Boston, for all their incredible and cheerful help.

PART I

CHAPTER 1

Introduction: the remarkable success story of the development of bortezomib*

On May 13, 2003, the US Food and Drug Administration (FDA) approved Velcade (bortezomib), a proteasome inhibitor (Fig. 1.1), under "Fast-Track" Application for the treatment of multiple myeloma, an incurable cancer of the blood that affects approximately 32,110 patients in the United States annually[1] (Box 1.1, Note 1). The story behind the development of this drug is quite unique and remarkable for many reasons, including:

- the initially fragile funding base of the original company that discovered and developed it (Myogenics/ProScript);
- the risks involved in pursuing a new molecular target (the proteasome) with a new, and allegedly ill-reputed, class of inhibitors (boronates);
- internal struggles and disagreements in the firm between the academic founders and the management regarding the indication for which the drug was to be developed;
- change of the original company's business model (from cachexia to inflammation to cancer) and subsequent change of the firm's name—from Myogenics to ProScript;
- enormously disruptive drop by Hoechst Marion Roussel, Inc., of proteasome inhibitors for inflammation and cancer and its return of its license rights on the drug to ProScript;
- change in leadership;
- depletion of initial funding and inability to secure further funding for clinical trials;
- the general lack of interest and total disbelief of the biopharmaceutical industry on bortezomib; and

* Adapted from Sánchez-Serrano, I., 2006. Success in translational research: lessons from the development of bortezomib. Nature Reviews Drug Discovery 5, 107–114.

The Core Model
ISBN 978-0-12-814293-6
https://doi.org/10.1016/B978-0-12-814293-6.00001-9

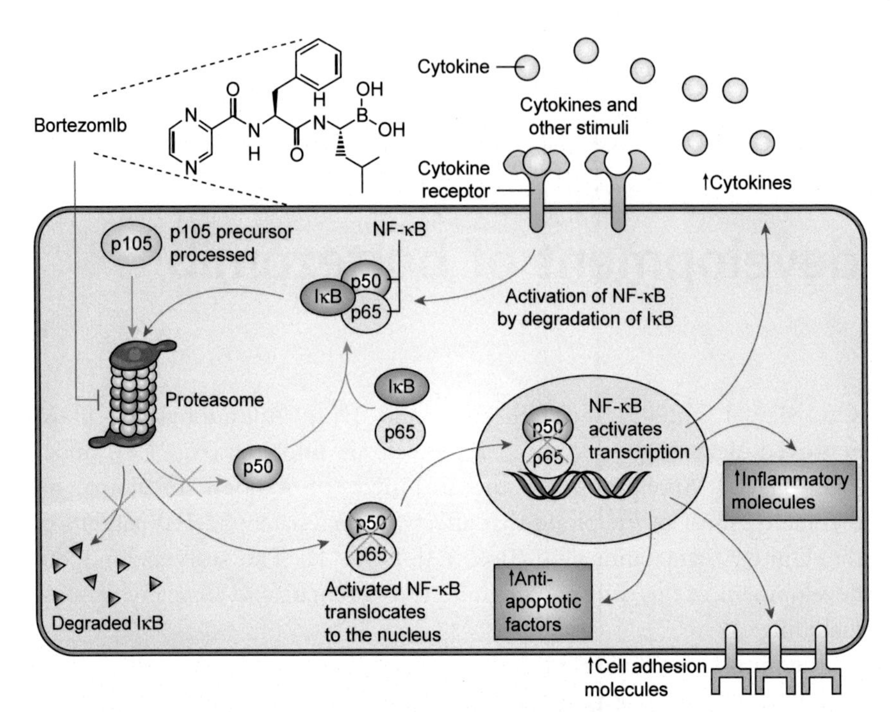

Figure 1.1 ***The proteasome, nuclear factor-κB, and bortezomib.*** The proteasome is a barrel-shaped multiprotein particle that destroys proteins that have been marked for degradation by conjugation to ubiquitin. Binding of the transcription factor nuclear factor-κB (NF-κB) to the inhibitor protein IκB in the cytoplasm renders NF-κB inactive. Cellular stimuli, such as cytokines, antigens, oxidants, viruses, and other agents, trigger a cascade of signal transduction events that phosphorylate and ubiquitinate IκB, leading to its degradation by the proteasome, which in turn liberates NF-κB for translocation into the nucleus. Once in the nucleus, NF-κB binds to the promoter regions of genes coding for proteins that are involved in the activation of transcription, growth, angiogenesis, anti-apoptotic factors, and cell-adhesion molecules. By inhibiting the proteasome, bortezomib inhibits the activation of NF-κB (orange crosses) and subsequent events that can promote tumor cell survival and proliferation. *(Source: Sánchez-Serrano, I, 2006. Success in translational research: lessons from the development of bortezomib. Nature Reviews Drug Discovery 5, 107–114.)*

- the resulting merging of ProScript into Cambridge-based LeukoSite and, six months later, the purchase of LeukoSite by Millennium Pharmaceuticals.

Despite all these major obstacles that could have quickly destroyed any company or drug development program, bortezomib managed not only to make it to the market but to do so in record time and at a significantly lower cost than biopharmaceutical industry average.

BOX 1.1 Additional notes

- Note 1: Multiple myeloma is predominantly a disease of the bone marrow and is the second-most common cancer of the blood (representing 1% of all cancers and 2% of all cancer deaths). It is estimated that, in the United States alone, approximately 120,000 people have multiple myeloma, and 32,110 patients develop the disease annually (18,130 in men and 13,980 in women). The average life expectancy of these patients is ~4.3 years, though some individuals can live up to 10–20 years with the disease (see the Multiple Myeloma Research Foundation website for further information: https://themmrf.org/).
- Note 2: The history of boronates as pharmaceutical agents is interesting. They were initially developed as serine–protease inhibitors by Dupont/ Merck. However, the original compound failed in Phase II clinical trials as a treatment for emphysema, and so boronates acquired a poor reputation among medicinal chemists. Adams's team linked a boronate group to Myogenics lead compounds, which showed considerably increased efficacy against the proteasome, and so created a novel chemotype.
- Note 3: Agreement reached for up to US$38 million from Hoechst Marion Roussel, Inc. (HMR), plus royalties paid to ProScript on sales of products deriving from this partnership; https://www.thepharmaletter.com/article/hmr-and-proscript-sign-r-d-pact; accessed 2/28/2019.
- Note 4: US$20 million in equity investment from Roche Group to ProScript, plus royalties on sales of products resulting from this collaboration; https://www.thepharmaletter.com/article/hmr-and-proscript-sign-r-d-pact; accessed 4/8/2019.
- Note 5: From a business point of view, it must be stressed that although the conclusion to focus on cancer was influenced by these results, bortezomib was first developed as an antiinflammatory agent and was licensed to HMR for that purpose (arthritis). The decision to pursue cancer as a business model was a fallback position because HMR chose to focus upstream of the proteasome in inflammation. Ultimately, HMR dropped proteasome inhibitors for inflammation and cancer, and returned its license rights on the drug to ProScript.

Why did bortezomib—which was considered doomed to failure from the beginning (Box 1.1, Note 2)—become a success story unlike countless other well-funded examples in the industry? How did it manage to reach the market in record time? What can we learn from this story and what relevance does it have in other related areas, such as the way in which the pharmaceutical industry develops novel drugs and the form in which governments craft and implement their innovation policies? What implications does this learning have for the academia–industry/public–private relationship in biopharma? What impact on global health care? Before answering these questions, I will narrate the remarkable story of the discovery and the development of this breakthrough drug.

Brief story of the discovery and development of bortezomib

In 1992, Alfred Goldberg, a cell biology professor at Harvard Medical School, decided to use the growing basic knowledge on the proteasome to create a biotechnology company focused on one goal: using inhibitors to block the proteasome. The proteasome is a hollow and cylindrical enzymatic complex that is present in both the cytoplasm and the nucleus of all eukaryotic cells and is necessary for the degradation of >80% of the cell's proteins.[2] This blockage was aimed at investigating the physiological roles of the proteasome and translating basic proteasome research into a therapeutic application. The company, Myogenics, was founded in 1993, and its objective was to target the ubiquitin–proteasome pathway to slow down the process of muscle wasting (cachexia) associated with fast protein degradation.

To create the company, Goldberg formed partnerships with two scientists at Harvard: Kenneth Rock, an immunologist and pathologist who had collaborated with Goldberg on studies of proteasome and antigen presentation; and Michael Rosenblatt, who brought a wealth of experience in drug development. Tom Maniatis, codiscoverer of the NF-κB pathway, who collaborated with Goldberg on gene transcription linked to the proteasome,[3] subsequently also joined.

In 1993, the company hired its first CEO, Frans Stassen, who came from Ciba–Geigy and had significant experience in drug development. A team of enzymologists (led by Ross Stein, who came from Merck) was also employed, and created the first inhibitors of the proteasome, which would eventually lead to bortezomib: peptide aldehyde analogs of the favored substrates of the proteasome's chymotrypsin-like active site.[4] These inhibitors (such as MG-132, which is still widely used in basic research) were distributed freely to many academic researchers. Next, Julian Adams (who was hired from Boehringer as Head Chemist, and who later became Executive Vice President of R&D) and the team of chemists that he led between 1994 and 1995 used a straightforward medicinal chemistry approach to create a dipeptide boronate, named MG-341, later known as bortezomib,[5] and other inhibitors (such as MG-519/PS-519).

An important characteristic of Myogenics was its extremely close collaboration with academia, in which important scientific knowledge on the proteasome and the pathways involved in inflammation and apoptosis was quickly created via productive partnership between several of the founders at Harvard and the scientists in the company. For example, the first inhibitors created by the company were immediately tested on cells in the laboratories of Goldberg and Rock, and this research afforded insights into the effect of proteasome inhibition on inflammation.

One of the initial key findings from the laboratories of Rock and Goldberg was that blocking the proteasome in vivo did not immediately alter the normal life of the cell.[3] Ensuing studies carried out in collaboration between Goldberg and Maniatis' laboratory at Harvard, such as the discovery by Vito Palombella and colleagues that the proteasome is very important in the activation of NF-κB[4], which is in turn involved in the inflammatory response,[6] led to a change of focus in the company. The company evolved from focusing on muscle wasting to inflammation (along with a corresponding change in company name, from Myogenics to ProScript). In August 1994, Avram Hershko suggested to Adams that the company investigate cancer as a potential disease target,[7] and investigations carried out by the company scientists showed that the proteasome inhibitors blocked the proliferation of cancer cells in vitro (bearing in mind the role of NF-κB in gene regulation). This strongly increased the interest in cancer, although the company continued to be focused on inflammation.

At this time, there was considerable tension between the founders, the company scientists, and the Scientific Advisory Board (SAB) owing to differences in opinion on the direction and strategies that the company should pursue: muscle wasting, inflammation, or cancer. A step forward in the cancer direction was the establishment of collaboration between ProScript and Beverly Teicher from the Dana Farber Cancer Institute (DFCI). Teicher was introduced to the company by Bruce Zetter (Harvard), who had been, in turn, introduced by Goldberg to the company as an SAB member. The group at ProScript then initiated their first proof-of-concept study in cancer in collaboration with Teicher in 1995. At this time, Teicher was studying angiogenesis and the role of toxic agents (such as alkylating agents) in cancer cells and provided ProScript with the first tumor mice models. By 1997, the group had shown that bortezomib (known as PS-341 at the time) inhibits tumor growth and metastasis in a mouse model of lung cancer; the results were published in 1999.[8] However, considerable skepticism existed about the drug based on what was considered its potential toxicity in humans.

In addition, Adams, through Maniatis, met David Livingston, a leading figure in the mechanistic field of oncology, and asked him to join ProScript's SAB. Livingston's role was crucial in steering the company to thought leaders in the cancer field, such as Kenneth Anderson from DFCI. At the end of 1995, ProScript established a collaboration agreement with Hoechst Marion Roussel (HMR) to develop orally active antiinflammatory and anticancer agents based on ProScript's ubiquitin–proteasome inhibition technology (see Box 1.1, Note 3). A year later, the company signed a drug discovery/development collaboration with Hoffmann–La Roche (Nippon Roche) on compounds to treat cachexia, muscle-wasting associated with cancer (see Box 1.1, Note 4). These important collaborations were secured by ProScript's second CEO, Richard Bagley, who was crucial not only in establishing these collaborations but also in further negotiations between ProScript and these companies, including the recovery of rights related to proteasome inhibitors from HMR.

Between 1996 and 1997, ProScript approached the National Cancer Institute (NCI) for collaboration. During this period, the NCI was interested in looking for new chemotherapeutic agents and had a large collection of cell lines in which PS-341 could be tested. This was a direct collaboration with Edward Sausville, Head of the Developmental Therapeutics Program (DTP) at NCI and Chair of the NCI Decision Network (the body that makes decisions on the commitment of NCI funds to new drug development initiatives arising either from NCI or from outside) and his team. After this collaboration and the initial data that resulted from it, the company started to focus increasingly on cancer, although it continued to pursue inflammation as well (see Box 1.1, Note 5).

Through 1997, ProScript continued to collect animal data and established collaborations with several academic researchers,[9] among them, Christopher Logothetis, Chairman of Genitourinary Oncology at the MD Anderson Cancer Center (MDACC). Logothetis was introduced to ProScript by David McConkey (MDACC) and was invaluable early on for the progression of ProScript's trials (P. Elliott, personal communication). Through Logothetis, ProScript met Howard Soule, Chief Science Officer at CaP Cure (now the Prostate Cancer Foundation).

After encouraging results from the NCI and animal studies carried out by the pharmacology group led by Elliott at ProScript[10], the company won unanimous approval and funding from the NCI, CaP Cure, and two academic groups, the Memorial Sloan Kettering Cancer Center (MSKCC) and the University of North Carolina (UNC) to conduct Phase I clinical trials. After successful initiation of the Phase I trial at MDACC, two additional trials were started. The first was at Memorial Sloan Kettering with David Spriggs and the second was with Robert Orlowski at UNC, who was introduced to the company by James Cusack and Albert Baldwin (UNC). Both trials were funded by the individual institutions, which shows that it is possible to find external funding for trials of promising drugs and thereby avoid the rapid depletion of the limited resources of a small biotechnology company. The trial at UNC focused on hematological malignancies and demonstrated (in 2000) that bortezomib was active in multiple myeloma[11] (see Box 1.2, Note 1).

Although bortezomib worked exceptionally well in animal models of inflammation, especially rheumatoid arthritis, ProScript realized that the therapeutic index was not large enough for chronic administration. Moreover, despite promising progress in cancer, such as the initiation of Phase I trials, ProScript's funds were almost depleted by June 1999. The company had received its initial funding from HealthCare Investment Corporation, as a leading investor (Dillon Read Venture Capital acted as coinvestors). ProScript was unable to secure subsequent funding for several reasons, including the pioneering nature of their technology, and that targeting the proteasomal apparatus with a drug that was considered to be too toxic was viewed as too risky, especially when taking into account the cost of the ensuing clinical trials. In addition, from a venture capital point of view, no suitable comparable drugs existed, the success for which could be used to provide support for taking further risks. The fact that HMR, during restructuring to form Aventis, dropped proteasome inhibitors exacerbated the negative feeling about further investing into the company.

Therefore, ProScript, like many other small biotechnology firms, fell victim to the financial market psychology of the moment and, faced by a funding shortage, reduced its staff and SAB. Adams, who had become bortezomib's champion, and the ProScript team made many efforts to promote the drug and establish collaborations with more than 50 companies, all of which declined the drug. Eventually, HealthCare Ventures decided to incorporate ProScript into another of their portfolio firms, Cambridge-based LeukoSite. The company team, together with some private investors, tried to buy the drug from HealthCare Ventures for

BOX 1.2 Additional notes

- Note 1: The actual timetable of events was as follows: National Cancer Institute (NCI) clinical candidate (Decision Network stage III; DN III) NCI approved funding for clinical trials on June 8, 1998. MD Anderson Cancer Center started first trial on October 7, 1998. Memorial Sloan Kettering Cancer Center started a second trial on February 15, 1999. New York University started a third trial (using NCI funding) on July 26, 1999. The University of North Carolina started fourth trial on November 8, 1999 (P. Elliott, personal communication).
- Note 2: Adams in an interview in *Forbes* (June 6, 2002): https://www.forbes.com/2002/06/06/0606mlnm.html#2e2e6b34164b; accessed 2/28/2019. Besides the DFCI, collaborations arose with other hospitals and institutions at the same time.
- Note 3: The example of bortezomib clearly indicates the value of a company's "translational" investment in assays relevant to clinical development. The scientists working for Peter Elliott and Julian Adams validated an assay for peripheral blood mononuclear cell proteasome activity and its modulation by bortezomib that crucially influenced the selection of the clinical schedule and allowed a tight dose-escalation scheme, leading to a highly efficient Phase I program. This should be highlighted as evidence favoring investment by drug companies in this type of assay, particularly for "first-in-class" agents such as bortezomib.
- Note 4: Millennium was granted "Fast Track Status" by the FDA in June 2002. Millennium filed a New Drug Application for bortezomib (Velcade; Millennium Pharmaceuticals) on January 21, 2003, under the provisions of Subpart H Accelerated Approval of New Drugs for Serious or Life Threatening Illnesses. On March 10, 2003, the FDA accepted the application and granted Priority Review Status. On May 13, 2003, bortezomib was approved under Fast-Track Status (http://investor.millennium.com/phoenix.zhtml?c=80159&p=irol-newsArticle&ID=411937&highlight=). Bortezomib generated revenues for Millennium of US$59.6 million in 2003, Millennium Annual Report 2003, Form 10-K, p1 (http://media.corporate-ir.net/media_files/irol/80/80159/reports/ar03.pdf). In 2004, bortezomib had net product sales of US$143 million, Millenium Annual Report 2004, Form 10-K, p1 (http://investor.millennium.com/phoenix.zhtml?c=80159&p=irol-newsArticle&ID=660798).

US$2.4 million in cash (Julian Adams, personal communication). Nevertheless, in July 1999, ProScript was sold to LeukoSite for US$2.7 million (approximately 187,000 newly issued shares of LeukoSite's common stock valued at $2.3 million and $430,000 in cash).[12]

Three months later, Millennium bought LeukoSite for US$635 million, because it was interested in LeukoSite's pipeline, especially CamPath (which also made it to the market), but had no interest in bortezomib.[13] Unsurprisingly, passing from company to company created major disruption to the bortezomib project. However, the ProScript team did not give up. Adams had regular meetings with Mark Levin, CEO at Millennium, to persuade him to keep the project alive and provide the necessary funding. In August 2000, the UNC clinical trial (that started in 1999[11]) demonstrated that bortezomib erased all signs of cancer from a 47-year-old woman, who months before was in the advanced stages of multiple myeloma. Given these data, Millennium decided to make bortezomib Millennium's most-funded drug (ML-341), and Adams and others from ProScript quickly assumed leadership roles at Millennium.

Adams' group decided to team up with Anderson, a multiple myeloma expert, and his group at the DFCI (specifically, Paul Richardson and Teru Hideshima) to conduct Phase II clinical trials at the institute. Consequently, an exchange of results and ideas took place back and forth between Millennium and Anderson and colleagues to discover more about the molecular mechanisms that made multiple myeloma more susceptible to bortezomib. At Millennium, David Schenkein, Dixie Esseltine, Barry Greene, Michael Kauffman, and others played important roles. This translational process was accomplished in the following way: basic research → clinical setting → patient feedback → return to basic research to gain further understanding of the molecular mechanisms involved (see Box 1.2, Notes 2, 3). In addition, through Anderson, Adams established contact with the Multiple Myeloma Research Foundation and the International Myeloma Fund (two advocacy groups that provide information and support to multiple myeloma patients and their relatives) that strongly supported the cause of bortezomib.

After the conclusion of Phase II clinical trials,[14] bortezomib was approved in record time in May 13, 2003, by FDA under Fast-Track Application (see Box 1.2, Note 4) as an injectable small molecule for the treatment of multiple myeloma. Millennium continued Phase III clinical trials and carried out the marketing of the drug.

Endnotes

1. https://www.cancer.org/cancer/multiple-myeloma/about/key-statistics.html; accessed 2/28/2019.
2. The history of proteasome research is far too extensive for this chapter; see the following reference for a description of the history and key players: Goldberg, A., 2004. In:

Adams, J. (Eds.), *Cancer Drug Discovery and Development: Proteasome Inhibitors in Cancer.* 17–38. Humana, Totowa.
3. Palombella, V.J., Rando, O.J., Goldberg, A.L., Maniatis, T., 1994. The ubiquitin-proteasome pathway is required for processing the NF-κB one precursor protein and the activation of NF-κB. Cell 78, 773–785.
4. Stein, R.L., Ma, Y.T., Brand, S., 1995. Inhibitors of the 26s proteolytic complex and the 20s proteasome contained therein. US Patent 5, 693, 617.
5. Adams, J. et al., 1995. Boronic ester and acid compounds. US Patent 5, 780, 554.
6. Barnes, P.J., Karin, M., 1997. Nuclear factor-κB: a pivotal transcription factor in chronic inflammatory diseases. New England Journal of Medicine. 336, 1066–1071.
7. Julian Adams, personal communication.
8. Teicher, B.A., Aran, G., Herbest, R., Palombella, V.J., Adams, J., 1999. The proteasome inhibitor PS-341 in cancer therapy. Clinical Cancer Research 5, 2638–2645.
9. Palombella, V.J. et al., 1998. Role of the proteasome and NF-κB in streptococcal cell wall-induced polyarthritis. Proceedings of the National Academy of Sciences of the United States of America 95 (15), 671–15, 676.
10. Adams, J. et al., 1999. Proteasome inhibitors: a novel class of potent and effective anti-tumor agents. Cancer Research 59, 2615–2622.
11. Orlowski, R. et al., 2002. Phase I trial of the proteasome inhibitor PS-341 in patients with refractory hematologic malignancies. Journal of Clinical Oncology 20, 4420–4427.
12. Bioline [online], 2005. https://scrip.pharmaintelligence.informa.com/deals/199910095; accessed 2/28/2019.
13. Langreth, R., Moukheiber, Z., 2003. Medical merlins. p.5 Forbes (23 June 2003).
14. Richardson, P.G. et al., 2003. A phase 2 study of bortezomib in relapsed, refractory myeloma. New England Journal of Medicine 348, 2609–2617.

CHAPTER 2

Lessons from the development of bortezomib: the Core Model

The way in which Myogenics/ProScript established several important collaborations with outside academics and agency/advocacy groups to move bortezomib forward was unusual, especially back in the early 2000s. Although some companies are now forming such kinds of collaborations in their programs, MyoGenics/ProScript did this exceptionally well and systematically, and so particularly benefited from these collaborations at critical points, when the company both needed scientific knowledge to move forward and lacked the necessary economic resources. The example of bortezomib emphasizes the potential power of maximizing such collaborative approaches and is useful in providing insights to policy makers, scientists, investors, and the public on how the process of drug development can be optimized.

Bortezomib, despite being a first-in-class drug that could have been shelved many times, managed not only to reach the market, but also to do so very quickly, in contrast to many well-funded efforts in industry. I propose that this can be explained using a "Core Model" (Fig. 2.1A,B), which defines and structures the roles and bidirectional interactions of the parties involved in the process of drug development.

In the development of bortezomib, this model could be described as follows (see Box 2.1 Note 1). The "Core" is represented by internal people and resources such as Goldberg, Maniatis, Rock, Rosenblatt, Adams, Elliott, Palombella, managers, other internal scientists, and private investors. The "Bridge" is represented by immediate collaborators, such as Teicher, Sausville, Logothetis, Soule, Spriggs, Orlowski, Anderson, Scientific Advisory Board (SAB) members, Millennium, etc.; and other private companies (via the simultaneous nonexclusive collaboration with external scientists or via non-exclusive SAB members who work simultaneously with these kinds of companies; see Box 2.1, Note 2). The "Periphery" is represented by the Prostate Cancer Foundation (CaP Cure), the Multiple Myeloma Research Foundation, the International Myeloma Fund,

The Core Model
ISBN 978-0-12-814293-6
https://doi.org/10.1016/B978-0-12-814293-6.00002-0

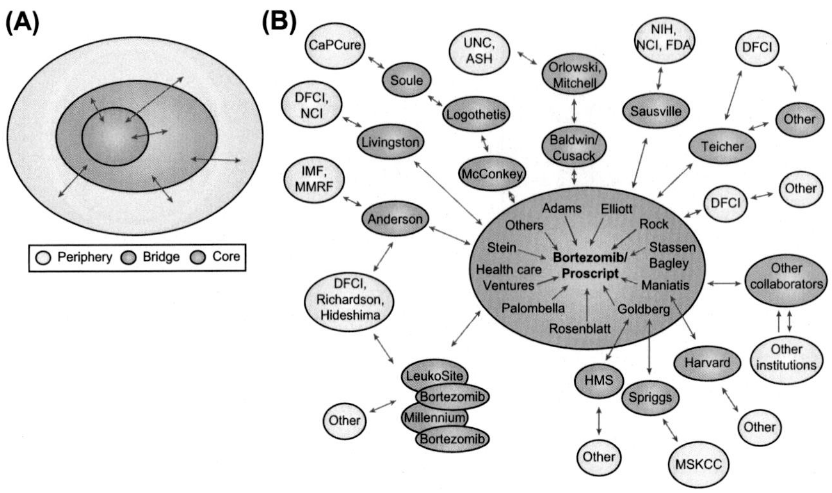

Figure 2.1 ***The Core Model.*** (A) The Universal Model. As an organizational paradigm, the Core Model consists of three major constituents: the "Core", the "Bridge", and the "Periphery". For instance, in a biotechnology start-up company, the Core represents the company's internal resources and people, who are hired because they have assets that are directly related to the Core's objective—making drugs. The Core needs a strong leader who is capable of keeping the enterprise focused and is able to secure collaboration with external people. The ideas of the Core are protected by patenting and secrecy. The Bridge represents the immediate collaborators and the private institutions to which the Core has indirect access through the external collaborators. The Bridge contains external scientists interested in similar problems or whose research would be greatly enriched because of the collaboration. It also includes consultants and Scientific Advisory Board (SAB) members (nonfounders) working in exclusive and nonexclusive ways. The Periphery contains the institutions/agencies interested in what the Core has to offer for the benefit of society, as well as the funding and regulatory structures that support the Core and the Bridge. The Periphery is an open, public, and cooperative system. The goal of the Core is to absorb this efficiently and legally as much relevant knowledge and information as possible. This knowledge and information come from its surroundings in three ways: first, via the leverage of the assets, professional backgrounds, and connections of the people within the Core; second, via the assets, connections, and expertise of the external collaborators within the Bridge; and third, via the support, relevant public knowledge, and know-how within the Periphery. The way in which this is accomplished is through knowledge transfer (collaboration established with external people to exchange assets), knowledge integration (incorporation and assimilation of external assets), and knowledge translation (conversion of all internal and external assets into a commercial therapeutic product). (B) Illustration of the roles of selected people involved in the development of bortezomib using the Core Model. (Reproduced from *Sánchez-Serrano, I, 2006. Success in translational research: lessons from the development of bortezomib. Nature Reviews Drug Discovery, 5, 107–114.)*

BOX 2.1 Additional notes

- Note 1: key people that interacted before and after the LeukoSite/Millennium acquisitions are included, because the "Core" never disintegrated, although legally ProScript had become part of LeukoSite and then Millennium, and several of its original members were no longer present. By the time ProScript was sold out, the company had already gained so much influence and momentum with its science and the establishment of a tight network interaction with the "Bridge" and the "Periphery", that even within Millennium it continued to exercise powerful "centripetal" influence. For example, the key Phase I clinical trial at the University of North Carolina, which showed safety and efficacy of bortezomib in humans, was planned and coordinated before the acquisitions took place. In addition, Phase II key people, such as Kenneth Anderson—who helped attract the Multiple Myeloma Research Foundation and the International Myeloma Fund—had already been contacted years before by Adams and colleagues, as suggested by David Livingston.
- Note 2: when collaborating with people in a nonexclusive way, parties may have a deep knowledge of what is happening in a specific field, but this is not intended to imply that there will be "leakage of information". By contrast, these collaborations could be extremely fruitful because they could help avoid unproductive paths and suggest new perspectives.
- Note 3: the company distributed bortezomib from early on, and the group has continued to do so, because the company did not have the resources to carry out all the studies, and high-profile academic groups gave the drug more credibility (P. Elliott, personal communication).

Memorial Sloan Kettering, Dana–Farber Cancer Institute (DFCI), National Cancer Institute (NCI), NIH, MD Anderson, University of North Carolina, FDA, etc (see Table 2.1 for a selected list of the key players involved and their assets).

Communications with the Periphery were established through individual people involved. These players traded assets (that is, materials, animal models, knowledge, and connections) and, in so doing, they advanced each other's research. Examination of their publications before, during, and after using the first-generation inhibitors (such as MG132 or lactacystin) and bortezomib reveals that the study of these drugs led to a better understanding of their mechanisms of action. The fact that the company decided to distribute bortezomib to outside researchers (see Box 2.1, Note 3), especially when it reached a high level of economic distress, was not risk free.

Table 2.1 Summary of the key players in the development of bortezomib (Velcade), their roles and contributions.[a]

Individual	Institution	Role	Assets
Bortezomib			
Alfred Goldberg	Harvard University, Dept. of Physiology	Founder, SAB	Muscle-wasting research, codiscovery of the proteasome
Tom Maniatis	Harvard University, Dept. of Biology	Cofounder, SAB	Research on IκB Kinase complex, NF-κβ. Experience at the Genetics Institute, SABs member, Editor of prestigious journals, introduced Vito Palombella
Kenneth Rock	Harvard University, DFCI	Cofounder, SAB	Cancer immunology and research on NF-κβ /Major Histocompatibility complex class I (MHC)
Michael Rosenblatt	Harvard University, BIMC	Cofounder, SAB	Blood cancer research and business experience at Merck leading drug development and bringing a drug to the market for the treatment of osteoporosis. Consultant to venture capital firms, SABs member
Richard Bagley	Myogenics/ProScript	CEO, SAB	Experience at BMS, SmithKline, CEO of ImmuLogic
Julian Adams	Myogenics/ProScript	Head of chemistry, SAB	Experience at Merck and Boehringer Ingelheim in drug discovery
Ross Stein	Myogenics/ProScript	Head of Biochemistry	Enzymology, experience from Merck
Vito Palombella	Myogenics/ProScript	Head of cell and molecular biology	Research on tumor necrosis factors (TNF), cancer immunology, interferon, NF-κβ. Maniatis' lab
Peter Elliott	Myogenics/ProScript	Head of Pharmacology Project leader	Pharmacology and experience in drug development from Alkermes, Inc. and the Glaxo group Ltd
David Livingston	Harvard University, DFCI	Collaborator, SAB	Cancer expert: Elucidation of pRB protein, work on BRCA1, BRCA2, BACH1, and BARD1. SAB chairman at NCI, other SABs member, EPA
Beverly Teicher	Harvard University, DFCI	Collaborator	Toxicology, angiogenesis, animal (mice) models, cell cultures, simultaneous collaboration with Immunogen, Genzyme, and Lilly
Edward Sausville	NCI	Collaborator	Head of the CTEP, NCI. NCI's scientific expertise and resources (in the form of cell lines and animal models), NIH/FDA

Velcade			
Christopher Logothetis	MD Anderson Cancer center, Texas	Collaborator	Head of Clinical Trials, MD Anderson, Texas. Founder of CaP cure. Leading expert in urogenital cancer
Howard Soule	CaP cure	Collaborator	Director of CaP cure, influence and weight on advocacy group, coordination of clinical research and trials, multiple connections with regulators, SABs member
Robert Orlowski	University of North Carolina–Chapel Hill	Collaborator	Interested in cancer and proteasome research. His father, Marian Orlowski, is codiscoverer of the proteasome. Connections with ASH
David Spriggs	Memorial Sloan Kettering Cancer center	Collaborator	MSKCC's chief of Developmental chemotherapy. Internationally recognized oncologist. Interested in solid tumors (gynecology) and translational research.
Kenneth Anderson	Harvard University, DFCI	Collaborator	Expertise in multiple myeloma, connections with the multiple myeloma Research Foundation and International Myeloma (BOD member)
Paul Richardson	Harvard University, DFCI	Collaborator	Expertise in multiple myeoloma, clinical trials aspect
Teru Hideshima	Harvard University, DFCI	Collaborator	Expertise in multiple myeloma, clinical research
Kathy Giusti	Multiple Myeoloma Research Foundation	Collaborator	President of MMRF. Multiple myeloma patient. Sales executive at Merck and GD Searle (Pfizer). Strong advocate to the multiple myeloma cause
Susie Novis	International Myeloma Fund	Collaborator	Head of IMF. Husband died of multiple myeloma

[a] *ASH*, American Society of Hematology; *BIMC*, Beth Israel Medical Center; *BMS*, Bristol–Myers Squibb; *BOD*, Board of Directors; *CEO*, Chief Executive Officer; *CSO*, Chief Scientific Officer; *CTEP*, Cancer Therapy Evaluation Program; *DFCI*, Dana–Farber Cancer Institute; *EPA*, Environmental Protection Agency; *FDA*, Food and Drug Administration; *IMF*, International Myeloma Fund; *MMRF*, Multiple Myeloma Research Foundation; *MSKCC*, Memorial Sloan Kettering Cancer Center; *NCI*, National Cancer Institute; *NIH*, National Institutes of Health; *SAB*, Scientific Advisory Board.

For example, the company took the risk that outside collaborators would have patented discoveries that, even if not valid or dominated by company patents, may have hindered the company and distracted resources. Although sharing the drug with outside scientists may have represented a tremendous loss of potential revenues for the company (if sold as a reagent), it opened crucial doors that eventually led to bortezomib reaching the market. Fig. 2.1B shows how these players interacted within the "Core Model" framework.

Although the science behind the development of bortezomib "worked", many people (including well-established companies) did not believe in bortezomib's commercial potential and rejected it, considering the drug to be too toxic, or the market too small. However, bortezomib is marketed today for several reasons. First, the Myogenics/ProScript "Core" had an outstanding, although small, scientific staff that made a number of highly influential basic discoveries, and distributed the initial inhibitors free to academic investigators, which led to rapid acquisition of knowledge in this area. Second, the collaborations established through the "Bridge" accelerated the generation of knowledge necessary for speedy approval. Third, once at Millennium, ProScript's team rapidly secured valuable resources from Millennium for additional clinical trials and marketing. Fourth, ProScript's Core used public resources within the "Periphery" very efficiently (for example, NCI, cancer advocacy groups, the hospitals carrying out the clinical trials, etc.). ProScript developed bortezomib through a "Core" modus operandi using knowledge transfer (collaboration established with external people to exchange assets), knowledge integration (incorporation and assimilation of external assets), and knowledge translation (the conversion of all, internal and external, assets into a commercial therapeutic product) (see Figs. 2.2 and 2.3).

Bortezomib would not have been successful in a large pharmaceutical company, as ProScript used a strategy that was very different from that taken by most of Big Pharma companies, which for the most part (back then and to some extent even today) were not centripetal from outside of their organizations (that is, they generally keep assets within the company). ProScript critically engaged and used the biology at every turn, made a convincing case for the drug, and so obtained support from the public sector at a critical juncture. The interest in bortezomib from the NCI was unusual: the NCI was going to pursue the molecule even if private funding was not forthcoming because of the drug's unique biological effects. Unlike many biotechnology start-ups or middle- or late-stage companies,

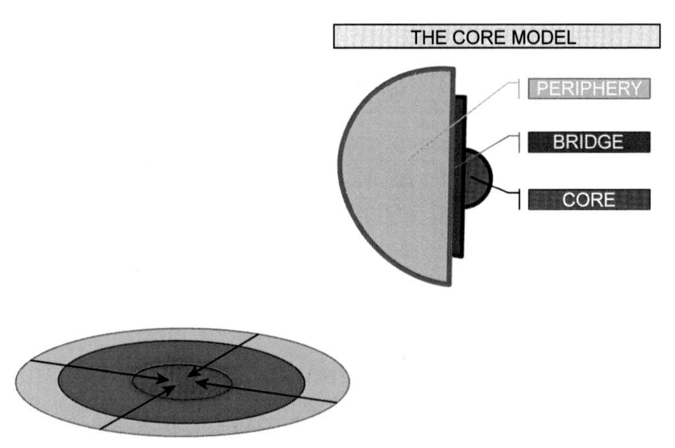

Figure 2.2 ***Core Model constituents.*** The Core and the Bridge are part of a bigger whole, which is the Periphery. The goal of the Core is to attract as many useful resources as possible from its surroundings.

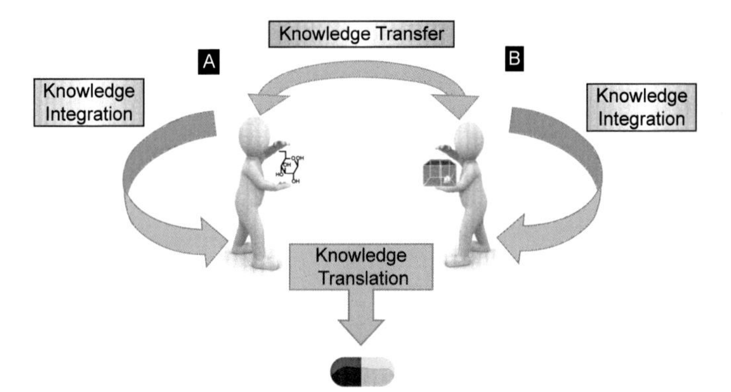

Figure 2.3 ***Core modus operandi.*** To reach its goals the "Core" uses the following modi operandi: knowledge transfer (collaboration established with external people to exchange assets), knowledge integration (incorporation and assimilation of external assets) and knowledge translation (the conversion of all, internal and external, assets into a commercial therapeutic product).

ProScript was not only focused on economic profits—it initially concentrated on a small and narrowly focused indication, rather than "holding out" for a more lucrative but scientifically less supportable indication. The company prominently engaged the patient advocacy community (CaP Cure) when starting the first Phase I clinical trials, and when the time came to recruit patients for definitive clinical trials, ProScript's "Core" was a key factor in involving the Multiple Myeloma Research Foundation and the International Myeloma Fund.

The Core Model and the academia–industry relationship

The relationship between academia and industry is generally understood in a unidirectional way, with basic science being translated into applied science (see Box 2.2). There is also the perception that in the academia–industry relationship, academia is exploited without receiving adequate benefits. However, upon close examination, it becomes clear that many scientists see translational research as a legitimate academic activity and that the academia–industry relationship is bidirectional. For example, academic research is stimulated by the questions that industry generates, which usually fall outside the scope, capabilities, and economic interests of the companies. Indeed, the process of academia producing ideas that are translated into commercial products is a cyclical one; academia provides answers to the questions created by new commercial products that, in turn, could lead to more commercial products. Although there is secrecy and proprietary

BOX 2.2 Basic, translational, and applied research

The definitions used to describe the biological research that is carried out in the domains of universities and companies vary widely. In this book, basic research is defined as the activity concerned primarily with the elucidation of the biological mechanisms and physicochemical processes of living organisms, irrespective of whether the findings of this research may eventually have potential therapeutic applications. The term translational research is used to refer to the translation of findings from the "bench to the bedside"; that is, translational research takes basic and preclinical findings and moves them into humans. Applied research is used to illustrate the development of commercial therapeutic applications, as related to health care interventions, with the objective of creating health, financial, and social benefits.

knowledge in the process of developing a drug, once the drug is marketed, the mechanisms involved in targeting the disease become public knowledge, resulting in more questions that could be investigated by academia. Considerations related to conflicts of interest and the pharmaceutical industry benefiting from public investment in research have generated a debate about the nature of the interactions between industry, academics, public institutions, researchers, and the propriety of such interactions. One of the important points emphasized by the "Core Model" is that these interactions are bidirectional in ways that benefit all parties and science in general, as well as providing society with new beneficial medicines. Such benefits are often missing from the debate (see Fig. 2.4).

In the case of bortezomib, scientists worked back and forth between academia and industry. The problems in the clinic gave rise to more research in the private sector and vice versa. This kind of interaction allowed the science to be integrally linked to the clinical studies, and, in turn, to allow the clinical studies to drive the science. Synergistic interaction has the potential, as illustrated here, to automatically correct directions that will not be productive. The transfer of knowledge between academia and industry allowed a better understanding of multiple myeloma, bortezomib's mode of action, and the mechanism by which the proteasome is related to other key pathways that regulate the cell cycle. Indeed, there has been an explosive interest in the proteasome and its role, not only in cancer but in other diseases as well. Many investigators are seeing bortezomib's antitumor effects in other types of cancer, and other agents created by Myogenics/ProScript are currently in clinical trials. Second-generation drugs have been

Figure 2.4 ***Implementation of the Core Model can have great health care and economic benefits for society.*** Through the Core Model and collaboration between the Core, the Bridge, and the Periphery, great health care benefits can be provided for society, saving time, labor, and investment.

developed in different laboratories and are awaiting funding to enter clinical trials. In fact, over time, Velcade became an over US $1 billion drug in the United States alone.[1]

In summary, academia has one characteristic that is very important to industry: it creates "full-stories" in terms of how living organisms work. Academia is a dynamic and open system that allows for rapid interchange of information among people from all over the world. This constant flow of people and ideas enriches scientific research and promotes progress. In other words, economic and social progress is achieved via a trade of assets and knowledge. From a broader perspective, the "Core Model' also has important implications for economic growth. It has been proposed as the state of technology that drives economic growth.[2] However, the present study on biomedical research is rooted in my firm conviction that trade (in this case of assets—knowledge and technology) and not the state of technology itself that drives economic growth. The "Core Model" explains how and why, and it could be generalized and used in other fields.

The development of bortezomib is an interesting case, because, despite ProScript starting without a drug, changing its business model, shifting in focus from muscle wasting to cancer, and running out of money, the company managed to access the right people and resources in a systematic way, leveraging cooperation with other, mostly public, institutions. Although there was intellectual property protection, the parties involved collaborated in a complementary, rather than competitive, way. ProScript's science was strong and pioneering, which is a standard requirement for any spin-off company, but "good science" is not enough to ensure success. The success of bortezomib was ultimately due to the tenacity of the people involved and the close collaboration, as explained in the Core Model (Box 2.1, Note 3), between academia, the private sector, private investors, public institutions, and advocacy groups. How many potential drugs like bortezomib have already been silently buried or are currently languishing? Policy initiatives in the areas highlighted above should help ensure that successes such as bortezomib become normal rather than exceptions.

In the ensuing three chapters, I will break down the Core Model into its three respective constituents: the Core, the Bridge, and the Periphery; and I will analyze each one of them vis-à-vis the current biopharmaceutical environment. This will allow us to answer the questions I posed at the beginning of the first chapter and contextualize the Core Model within a broader health care and socioeconomic framework.

Endnotes

1. Wolfe, J., 2017. U.S. Court Upholds Takeda Patent on Cancer Drug Velcade, Business Insider, July 17, 2017. Available from: https://www.businessinsider.com/r-us-court-upholds-takeda-patent-on-cancer-drug-velcade-2017-7.
2. Solow, R.M., 1956. A contribution to the theory of economic growth. Quarterly Journal of Economics 70, 65−94.

PART II

CHAPTER 3

The "Core": when innovation meets leadership

It is often said that the success and survival of the biopharmaceutical industry depend on its ability to innovate, in other words, to bring new drugs, new medical treatments, or new medically related products to the market to satisfy unmet medical needs. The more new pharmaceutical products sold, the higher the profits for the industry and, hopefully, the greater the benefit for society.

The innovation that we see in this industry is both the end and the beginning of a very lengthy, complex, and expensive process that takes place in both academia (public sector) and industry (private sector). Innovation in biopharma can come as a direct result of the application of basic academic research, which is hypothesis testing–driven; it can appear empirically, via observation and experience; heuristically, through trial and error; serendipitously, by mere chance; or it can be "directed," that is, "designed" to meet market demands, or to sustain growth, enhance competitiveness, and ensure commercial survival. In general, innovation in this industry, as in any other, can encompass many things depending on the context: from a new idea, device, or method to the application of better solutions that meet new requirements, unarticulated needs—i.e., those that consumers do not know that exist or are not aware that exist—or existing market needs. But it can also include "disruption"—a concept in which innovation creates a new market and value network that eventually break away from or "disrupt" an existing market and value network, displacing established market-leading firms, products, and alliances.[1]

In the biopharmaceutical world, in praxis, innovation is actually a conflation of three distinct yet closely interrelated concepts: novelty, discovery, and invention. Novelty refers to something that is new; discovery, to *finding* something new; invention refers to *creating* something new that did not exist. They are all forms of *knowledge*.

Therefore, although it is true that the biopharmaceutical industry depends on innovation to survive, innovation in itself is not enough to drive

The Core Model
ISBN 978-0-12-814293-6
https://doi.org/10.1016/B978-0-12-814293-6.00003-2

success. Innovation, as a form of knowledge, needs to be developed or incorporated adequately and effectively (in other words, "tailored") into the company's main mission, capabilities, resources, and culture, and, then, be translated, according to market opportunities and development costs (among other factors), into high-quality products for health care benefits of patients. This depends on the managerial skills, competence, and vision of the leaders within the company and how well they are able to understand simultaneously costs, nature of the innovation, technological challenges, market opportunities and needs, future of the industry, strategic partnerships and alliances, finances, timing, actual uses and benefits of the technology/innovation, etc. In addition, they must also be able to deploy the company's resources, such as innovation and *human capital*, in the best possible manner through *organization*. For these reasons, leadership and people are as important as, or even more important than innovation per se in the survival of a biopharmaceutical enterprise. Innovation, whether directed or serendipitous, is created, identified, discerned, assessed, and utilized by people. Thus, to survive, the biopharmaceutical industry depends primarily on its people and *then* on innovation. It is well-documented[2,3] that innumerable companies in the biotech and pharmaceutical fields, as well as other industries, have failed big–time, despite having the finest technology and innovation. This has been due to poor management of knowledge, wrong managerial decisions, inadequate organization, ineffective funding, lack of sufficient market, failed partnerships, or even the inability to clearly identify and define the best application, the best market opportunity, the best timing—among other reasons—all of which are contingent on leadership. A biopharmaceutical firm will only be as good and innovative as its leadership.

The biopharmaceutical industry and global health care

Two of the major challenges that the biopharmaceutical industry has to confront today, which, by the way, have a direct impact in the survival of the world's health care systems, are: first, the high cost of research and development (including marketing); and second, the high attrition rate of developing compounds. It is now estimated that new drug development can cost over US$2 billion and take between 10 and 15 years to accomplish.[4] For every drug that ultimately receives approval from the US Food and Drug Administration, some 5000 to 10,000 compounds do not make it through the process[5] (see Fig. 3.1 and 3.2 for a schematic diagram of the

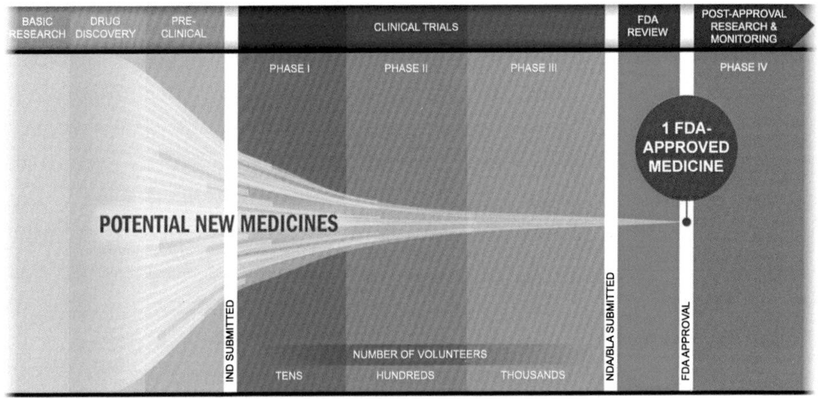

Figure 3.1 ***Drug Discovery and development is a lengthy and expensive process. It can cost over US$2 billion and 10–15 years to bring a novel drug to the market.*** *(From https://www.phrma.org/graphic/the-biopharmaceutical-research-and-development-process (original source).)*

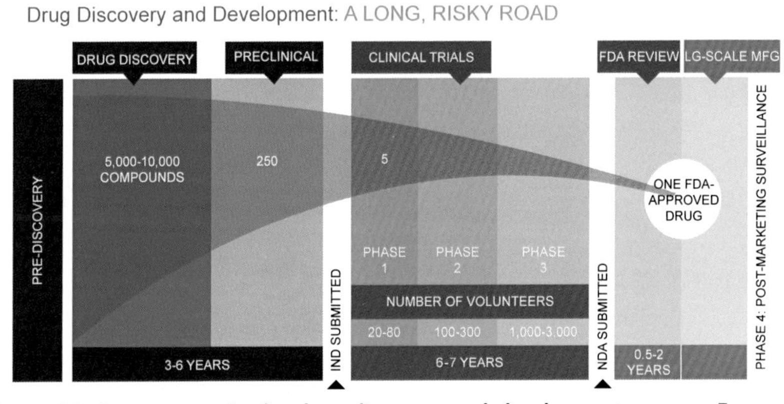

Figure 3.2 ***Success rate in the drug discovery and development process.*** For every drug that ultimately receives approval from the US Food and Drug Administration, some 5000 to 10,000 compounds do not make it through the process. *(From Pharmaceutical Research and Manufacturers of America (PhRMA) (original source).)*

drug discovery and development process and the success rates). The likelihood of a drug at Phase I making it all the way to the market has recently been estimated at 14%.[6] Although these statistics are always a source of contention, the truth is that it is very difficult to develop a new drug. In fact, since the 1950–1960s, we have observed a sharp decline in the approval of New Molecular Entities (NMEs), whereas the investment in R&D has grown exponentially (Fig. 3.3). This can be due to a multitude of factors, including the realization of the high complexity of biological systems and, therefore, a higher complexity of the science; the development of technological tools that are more novel; and the advent of genomics and sequencing of the human genome, which have provided a very large number of highly novel, but risky targets.[7] For instance, the human genome has ~3000 druggable targets.[8] Other factors include the necessity of more highly skilled researchers; perhaps higher regulatory hurdles; and the number of large clinical trials being performed at the same time by any given pharmaceutical company; etc. As I have discussed in *The World's Health Care Crisis*,[9] all of this has a direct impact on how drugs are priced and whom they are going to benefit. In the United States, which is the world's leader in biopharmaceutical innovation,[10,11] drugs are priced according to whatever the market can bear, because there are no official price controls.[12] Although the rationale behind lack of price controls for newly developed drugs is the fostering of innovation, this policy also comes with several curses. It not only has a negative impact on the economy of the

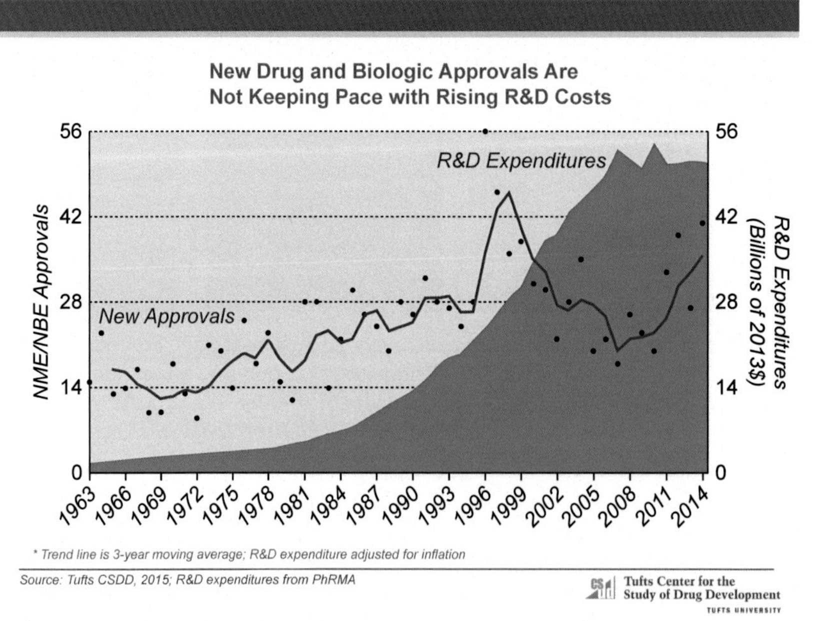

Figure 3.3 New drug and biologic approvals are not keeping pace with rising R&D costs. *(Reprinted with permission.)*

US health care system in terms of the portion that it consumes from the total health care expenditure (approximately 10%[13]) and from the individual's pocket (taxes, out-of-pocket expenditures, health insurance premiums, etc.), but it also creates negative spillover effects on the rest of the world.[14] This is especially true in the developing nations and industrialized countries with tightly regulated drug price controls, as is the case of most European countries, Australia, Japan, some Latin American countries, and Canada, among many others.

A global health care system, in which access to newly developed drugs increasingly becomes a privilege for the rich, is unsustainable and doomed to implode. (See Table 3.1 for a list of the price of some of the most expensive drugs recently developed.) Therefore, to keep a novel drug accessible to the largest number of patients possible in the world, we need to come up with new models for R&D that will enable lower costs of production, higher productivity, higher quality, and lower pricing. These models should be implemented at the "micro" and the "macro" levels. The micro level refers to the way in which basic, applied, and translational research are performed in academia and industry with the clear objective of better understanding the biological causes and mechanisms of diseases and to find treatment and cures for them. The macro level refers to the organization of the firm, its strategies, and the way in which it deploys its human capital to make the most of its limited resources.

The implementation of these models necessitates from all sectors involved in drug discovery and development the following:

1. An understanding of the importance of basic research and innovation in drug discovery and development;
2. An understanding of the mechanisms of translation of these basic research efforts and innovation into pharmaceutical and health care applications or products;
3. An understanding of the roles of leadership, people, and institutions in this process;
4. A desire to utilize in an optimal manner all the resources available in society to improve drug discovery and development with the goal of making pharmaceutical drugs more accessible and cheaper to society;
5. An ethical attitude from the pharmaceutical industry about its quintessential role in creating medications that save and improve the lives of millions of people around the planet.

Table 3.1 The most expensive specialty Drugs (2017).

Drug name	Company	Indication	Price/Year
1. Glybera (alipogene tiparvovec)	uniQure	Lipoprotein lipase deficiency	$1.2 million
2. Ravicti (glycerol phenylbutyrate)	Horizon pharma	Cycle disorder	$793,000
3. Brineura (cerliponase alfa)	BioMarin pharmaceutical	Late infantile neuronal ceroid lipofuscinosis type 2 (CLN2) disease, a form of Batten disease	$700,000
4. Carbaglu (carglumic acid)	Orphan Europe	N-acetylglutamate synthase deficiency	$419,000–$790,000
5. Lumizyme (alglucosidase alfa)	Genzyme	Pompe disease.	$524,000–626,000
6. Actimmune (interferon gamma-1b)	Horizon	Life-threatening osteopetrosis	$244,000–572,000
7. Soliris (eculizumab)	Alexion	Hemolytic uremic syndrome (aHUS)/paroxysmal nocturnal hemoglobinuria (PNH)	$432,000–$542,000
8. Folotyn (pralatrexate)	Allos therapeutics	Relapsed or refractory peripheral T-cell lymphoma (PTCL)	$96,000–$472,000
9. Demser, (metyrosine)	Valeant pharmaceuticals	Pheochromocytoma	$96,000–$472,000
10. Ilaris (canakinumab)	Novartis	Periodic Fever syndromes (cryopyrin-associated periodic syndromes, tumor necrosis Factor receptor associated periodic syndrome, Hyperimmunoglobulin D syndrome/Mevalonate Kinase deficiency, familial mediterranean Fever) Active systemic juvenile idiopathic arthritis	$379,000–$462,000

From Drugs.com; https://www.drugs.com/slideshow/top-10-most-expensive-drugs-1274; October 2017 *(original source)*.

The Core Model is one such model. Let us discuss its basic constituent, the "Core." in this chapter to explore how it can impact biopharmaceutical innovation and global health care.

The "Core"

For a moment, let us explore this concept stepping outside the biopharmaceutical world and entering the world of linguistics and economics. The word "core," in English, derives from the Latin word "cor," via the French "coeur." It means both "heart" and "nucleus." According to the Merriam-Webster's Dictionary, "core" is "a central and often foundational part usually distinct from the enveloping part by a difference in nature." The dictionary provides another definition for "core" as "a basic, essential, or enduring part (as of an individual, a class, or an entity)."[15] All these definitions are apposite to the definition of the Core within the Core Model, which in its broadest sense I define as "a self-contained (autonomous) entity that is trying to reach a particular goal." A Core can be a single individual, a group of organized individuals (an association), a firm, a foundation, a not-for-profit organization, a governmental institution, a church, a start-up company, or a giant corporation—with a goal or set of goals.

From an economic perspective, the Core can be related to, but is different from, the microeconomic concept of the "firm." According to the "theory of the firm,"[16,17] founded in neoclassical economics, firms (including businesses and corporations) exist and make decisions in a variety of areas such as resource allocation, production technique, pricing adjustments, and quantity produced to maximize net profits. Firms interact with the market to determine pricing and demand and then allocate resources according to models that look to maximize net profits, among other characteristics. Although all of this can be true for a Core, the essence of this concept goes far beyond the "firm." Let us delve into this as it is tantamount for our model.

To begin with, the goal of a Core is not necessarily to maximize net profits, but to achieve a particular goal or set of goals regardless of the nature of that goal. For instance, an individual who wants to meet the President of his/her nation is as much of a Core as a start-up company, such as Myogenics, or another firm, wishing to develop a new anticancer drug, or a large corporation that specializes in selling soft drinks. The Core can be a legally registered company or not. It can have Intellectual Property Rights and secrecy or not; it can have a product to sell or not; it may have tangible

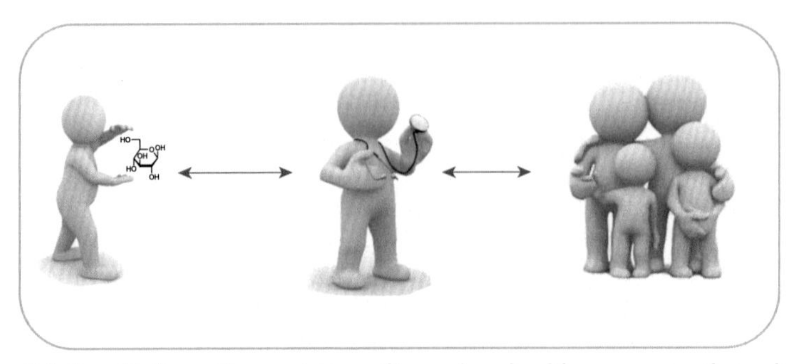

Figure 3.4 *Benefits from other social* benefits, such as health care, come through other individuals and institutions in society in a bi-directional interaction.

assets (such as products or real property) or intangible assets (such as knowledge). However, all Cores share one inherent characteristic: they all have at least one asset. And an asset here could be anything: from a single personal relationship or connection, to a personal talent or qualification, to an extraordinary wealth of knowledge, to a first-in-class compound, to an incredible amount of money, etc. Because we, as humans, live in society and interact with other people on a daily basis, every single individual on the planet has at least a personal relationship that, together with one's own personal knowledge and qualifications, is an asset.

The way in which the Core achieves its goals is by capitalizing on its assets, by bartering them, exchanging them, trading them, sharing them, selling them, or even donating them to other (external) players, which I have already defined as "Bridge," via a mechanism of assets transfer, integration, and translation (Fig. 2.3). I will elaborate on the Bridge and its importance in the next chapter. At this point I need to mention that no matter how self-centered or egoist our goal may be, because we live in society and are interdependent, our goals and actions will always involve the "other" and society, in one way or another. It is then in relationship with others that we not only create goals (and goods and services), but also have the desire to realize them, and in fact it is through others that we realize them (See Fig. 3.4).

How does it all relate to biopharma and to global health care?

The Core and the pharmaceutical industry

The chemical and pharmaceutical industry started, indirectly, with a Core: the British chemist William Henry Perkin (1838–1907). In 1856, Perkin, stimulated by his mentor, the German chemist August von Hoffmann, tried to synthesize (i.e., create artificially) quinine, a drug that had already been isolated in 1820 by Pierre Pelletier and Joseph Bienaimé Caventou from cinchona bark. Quinine, the use of which by South American Indians was recorded around 1630 by Spanish conquistadors, was in great demand in Europe for the treatment of malaria and other fevers.[18] However, cinchona bark grows in mountainous regions of South America, and at that time, the supplies of this plant in Europe were very scarce while the demand was high. Given the progress that had taken place in synthetic chemistry, it was thought that this substance, the chemical formula for which was known ($C_{20}H_{24}N_2O_2$), could be produced synthetically in a laboratory or chemical factory. Therefore, great commercial and economic incentive existed to pursue this idea.

Perkin tried to synthesize quinine by coal tar distillation, because, through this method, it is possible to obtain several intermediate compounds such as benzene, toluene, naphthalene, and anthracene—to mention only four important substances—which are the starting point for countless other products.[19] But to his own surprise, Perkin initially obtained a dark precipitate that was certainly not quinine. After subsequent distillations, he managed to extract from benzene a brilliant purple dye, subsequently named "mauveine" or "aniline blue." From aniline, Perkin obtained the first artificial dyestuff ever produced. Single-handedly, at 18 years old, he secured funding and external collaborations and went on to create the modern synthetic dyestuffs industry, introducing a new range of colors used for many applications such as in photography, in which Perkin was very interested, and microscopy. In the latter, it was used to stain specimens and reveal structures that otherwise would have been invisible to the human eye. His manufacture of dyestuffs was a great commercial success.

Given the numerous types of compounds that could be derived from coal tar, parts of this industry widened their range of interest to other fine chemicals. For instance, dyes gave rise to the manufacture of sulfuric and nitric acids and caustic soda; in turn, these led to the creation of artificial fertilizers, explosives, and chlorine. They could also give rise to medicines because coal tar contained many of the aromatic and aliphatic building blocks that became the tool kit of medicinal chemistry. After a series of distillations, chemical derivatives similar to the ones already characterized in some medicines could be obtained, which was exactly what Perkin had in mind when he attempted to synthesize quinine from coal tar. Most pharmaceutical companies in the second half of the 19th century in

Germany (Hoechst, Agfa, Bayer), Switzerland (Ciba, Geigy, Sandoz, F. Hoffmann–La Roche), England (Burroughs Wellcome, Glaxo, Beecham Group), the United States (Eli Lilly, Abbott Laboratories, SmithKline, Bristol-Myers, Upjohn, Pfizer), and in other countries started in similar "Core" ways, as I have already illustrated in *The World's Health Care Crisis*[20]. Likewise, the biotech industry in the United States (Genentech, Genzyme, Biogen, Amgen, Genetics Institute, etc.) in the early- to mid-1970s, started in similar fashion.[21]

As indicated previously, a Core can be a not-for-profit entity or fully for profit. For now, for the sake of clarity, given that here we are dealing with the biopharmaceutical industry and global health care, let us explore the Core within the context of profitability, intellectual property, and limited resources—financial and otherwise—which are a basic economic rule. Let us also return to the case of Myogenics/ProScript for the development of bortezomib, because this is a quintessential example. (We will discuss Big Pharma in Chapter 7)

In the previous chapter, I defined this type of "for-profit" Core as a given company's internal resources and people, who are hired because they have assets (knowledge, connections, experience, etc.) that are directly related to the *goal* of the Core—in this case, making drugs to treat diseases. An important characteristic of the Core is having a strong leader who is capable of keeping the enterprise focused and is able to secure collaboration with external people to fulfill its mission. The ideas of the Core are, in this case, protected by patenting and secrecy. This definition highlights several important things: crucially, adequate management of technology/innovation and human capital.

The starting point of any biotechnology start-up focused in human medicine is a breakthrough technology (innovation) that promises an important direct application in the treatment of a given human disease. In general, though not exclusively,[22] innovation arises in academic research centers. These centers license out novel technologies to small, medium, and large firms created by entrepreneurs (in exchange for milestone payments, royalties, or equity) for further development. Through seed funding, the CEO or leader of the company develops a strategy for development of the novel technology and forms a team that will move the technology forward. He or she will secure adequate funding for the company and the studies that need to be performed through preclinical experiments (e.g., toxicology, pharmacodynamics, pharmacokinetics, blood–brain barrier permeability, etc.) and clinical trials (safety and efficacy). In normal scenarios, it is expected that the technology will have a very strong Intellectual Property (IP) position. Over time, the company will create value and generate further assets, including further intellectual property.

The assets that a Core firm may have in its earlier stages can be tangible (such as specific molecules) or intangible (for instance, personal connections, an understanding of the mechanisms involved in a given biochemical pathway, or a particular know-how). The case study of Myogenics/ProScript provides us with useful examples of several important assets that the firm can leverage for further growth in the presence of small funding:

- Firsthand (and novel) knowledge on the proteasome (intangible asset)
- First-in-class drug target, the proteasome (tangible asset)
- High-quality research at a very prestigious university, Harvard Medical School (intangible asset)
- First inhibitors of the proteasome; including boronates (tangible asset)
- Very expert team, with great experience from academia and industry and excellent personal connections (intangible asset)
- Support from academic institutions, including Harvard (tangible asset in terms of infrastructure, equipment, basic research funding, and personnel; intangible asset if considered as connections, reputation, and access to other scientists interested in similar problems).
- Leadership/championship (intangible asset)
- Willingness to collaborate with external groups (intangible asset)
- Motivation and belief in the enterprise (intangible asset).
- Intellectual property (tangible asset; see Fig. 3.5)
- Research results, knowledge (intangible asset).
- Initial seed funding (tangible asset).

Human capital played, as an asset, a determinant role in the eventual realization of Myogenics/ProScript's goals. At the risk of repeating what we said in the previous chapters, let us analyze how the internal human resources of Myogenics/ProScript were capitalized in the most effective way:

Alfred Goldberg, from Harvard Medical School, throughout his academic life, was interested in the proteasome, a subject in which he was one of the pioneering figures. One of his goals, in creating a company, was to learn more about the effects that proteasome inhibitors would have in the cell, in vivo. Prior to the founding of Myogenics, Goldberg had established collaboration with a large number of academic labs, with industry as well as with the NIH. For Myogenics, Goldberg collaborated first with Kenneth Rock, from the Dana–Farber Cancer Institute (DFCI; Affiliated to the Harvard Medical School), who brought in his background

US005780454A

United States Patent [19]
Adams et al.

[11] **Patent Number:** **5,780,454**
[45] **Date of Patent:** **Jul. 14, 1998**

[54] **BORONIC ESTER AND ACID COMPOUNDS**

[75] Inventors: **Julian Adams,** Brookline; **Yu-Ting Ma.** Needham: **Ross Stein.** Sudbury: **Matthew Baevsky,** Jamaica Plains: Louis Grenier; **Louis Plamondon.** both of Belmont, all of Mass.

[73] Assignee: **ProScript, Inc.,** Cambridge. **Mass.**

[21] Appl. No.: **549,318**

[22] Filed: **Oct. 27, 1995**

Related U.S. Application Data

[63] Continuation-in-part of Ser. No. 442,581. May 16. 1995. which is a continuation-in-part of Ser. No. 330.525. Oct. 28. 1994. abandoned.

[51] **Int. Cl.**[6] **C07F 5/02;** C07F 5/04: A61K 31/69

[52] **U.S. Cl.** **S14/64;** 544/229

[58] **Field of Search** 544/229; 514/64

[56] **References Cited**

U.S. PATENT DOCUMENTS

5.550,262 8/1996 Iqbal et al.,
5.614,649 3/1997 Iqbal et al.,

FOREIGN PATENT DOCUMENTS

0 471 651 2/1992 European Pat. Off.,
WO 92/07869 5/1992 WIPO .
WO 93/21213 10/1993 WIPO .
WO 93/21214 10/1993 WIPO .
WO 94/21668 9/1994 WIPO .
95/09858 4/1995 WIPO .

OTHER PUBLICATIONS

Bachovin. W.W., et al., "Nitrogen–15 NMR Spectroscopy of the Catalytic–Triad Histidine of a Serine Protease in peptide Boronic Acid Inhibitor Complexes." *Biochemistry* 27:7689–7697 (1988).

Berry. S.C., et al., "Interaction of Peptide Boronic Acids With Elastase: Circular Dichroism Studies." *Proteins: Structure, Function, and Genetics* 4:205–210 (1988).

Kettner. C.A., et al., "Kinetic Properties of the Binding of a-Lytic Protease to Peptide Boronic Acids." *Biochemistry* 27:7682–7688 (1988).

Kinder. D.H., and Katzenellenbogen. J.A., "Acylamino Boronic Acids and Difluoroborane Analogues of Amino Acids: Potent Inhibitors of Chymotrypsin and Elastase." *J. Med. Chem.* 28:1917–1925 (1985).

Kinder. D.H., et al., "Antimetastatic Activity of Boro–Amino Acid Analog Protease Inhibitors against B16BL6 Melanoma in vivo." *Invasion Metastasis* 12:309–319 (1992).

Lim. M.S.L., et al., "The Solution Conformation of (D)Phe–Pro–Containing Peptides: Implications on the Activity of Ac–(D)Ph–Pro–boroArg–OH. a Potent Thrombinn Inhibitor." *J. Med. Chem.* 36(13): 1831–1838 (Jun. 25. 1993).

Matteson. D.S., et al., "(R)–1–Acetamido–2–phenylethaneboronic Acid. A Specific Transition–State. Analogue for Chymotrypsin." *J. Am. Chem. Soc.* 103:5241–5242 (1981).

Takahashi. I.,H., et al., "Crystallographic Analysis of the Inhibition of Porcine Pancreatic Elastase by a Peptidyl Boronic Acid: Structure of a Reaction Intermediate." *Biochemistry* 28:7610–7617 (1989).

Tsai. D.J.S., et al., "Diastereoselection in Reactions of Pinanediol Dichloromethaneboronate." *Organometallics* 2:1543–1545 (1983).

Tsilikounas. E., et al., "Identification of Serine and Histidine Adducts in Complexes of Trypsin and Trypsinogen with peptide and Nonpeptide Boronic Acid Inhibitors by ^{1}H NMR Spectroscopy." *Biochemistry* 31:12839–12846 (1992).

Veale. C.A., et al., "Nonpeptidic Inhibitors of Human Leukocyte Elastase. 5. Design. Synthesis. and X–ray Crystallography of a Series of Orally Active 5–Aminpyrimidin–6–one–Containing Trifluoromethyl Ketones." *J. Med. Chem.* 38(1):98–108 (Jan. 6. 1995).

Primary Examiner—Robert W. Ramsuer
Attomey, Agent, or Firm—Sterne, Kessler, Goldstein & Fox. PLLC

[57] **ABSTRACT**

Disclosed herein is a method for reducing the rate of degradation of proteins in an animal comprising contacting cells of the animal with certain boronic ester and acid compounds. Also disclosed herein are novel boronic ester and acid compounds, their synthesis and uses.

22 Claims, 3 Drawing Sheets

Figure 3.5 Patent application for bortezomib. *(From United States Trade and Patent Office (USTPO) (original source).)*

cancer immunology and how nuclear factor (NF)-κB was involved in antigen presentation. Rock's competence in immunology and cancer research was matched by his connections at the DFCI, Harvard Medical School, the field of oncology, and many other academic and industrial institutions.

Then, Goldberg and Rock engaged Michael Rosenblatt, who brought with him his expertise in blood cancer, but more importantly his prestige in the medical and academic worlds, as he had been a Professor at the Harvard Medical School, Director of the MIT—Harvard Health Sciences Program, among other positions. Furthermore, he had great experience in the pharmaceutical industry (working at Merck for 10 years leading drug development and bringing a drug to the market for the treatment of osteoporosis). He also had business experience as a consultant to venture capital firms and had been a member of many Scientific Advisory Boards.

Tom Maniatis, who joined the company as a late founder, brought in his expertise in the interferon gene and IκB Kinase Complex research, his connections at Harvard, his experience and connections at Cold Spring Harbor Laboratory (one of the most outstanding research centers in the world), as well as his thorough knowledge on NF-κB, of which he is a codiscoverer. Furthermore, from a business perspective, Maniatis had been involved in prior years with the biotech firm Genetic Institute and belonged to the Scientific Advisory Board of Ariad Pharmaceuticals (Cambridge, Massachusetts), among others, which allowed him to contribute greatly with his experience on different aspects of cancer research. He was also editor of leading scientific journals, such as *Cell*, and member of many academic boards. Very importantly, Maniatis brought with him Vito Palombella, a postdoctoral fellow, who graduated from the New York University and who brought his expertise in tumor necrosis factors (TNF) and cancer immunology.

Once the company was formed, through Goldberg's initiative, it brought the first CEO, Franz Stassen with extensive experience in drug development at Ciba—Geigy. Stassen hired Ross Stein as head enzymologist, who had great experience from Merck, and Julian Adams, as head chemist. Adams, at the moment of joining the company, brought his years of experience at Merck and, more importantly, his knowledge on drug discovery learned at Boehringer Ingelheim. There, Adams was able to find drug candidates from a vast library of compounds and developed the drug Viramune (nevirapine) for human immunodeficiency virus (HIV). His scientific ideas led him to connections with the regulatory entities and the different institutions/mechanisms involved in bringing a drug to the market. Peter Elliott, Project Leader, brought his expertise in pharmacology and experience in drug development from Alkermes, Inc., and the Glaxo Group Ltd. The second CEO of the company, Richard Bagley, with experience at Bristol—Meyers Squibb, GlaxoSmithKline, and Immulogic Pharmaceutical Corp., brought important skills in establishing partnerships and negotiations with Big Pharma.

The competence, experience, and, in particular, the personal connections that each one of these individuals had prior to joining Myogenics/ProScript proved extremely important in the success of bortezomib, because these people were able eventually to bring in other people. One such, David Livingston, came from DFCI, an important and well connected thought leader in oncology, who joined the company as a Scientific Advisory Board Member, and through whose personal connections the company had many other important doors open to them. Selecting very well the people that one works with can have an enormous impact on the type of knowledge that one acquires. These people contribute to the Core (or new company) with the knowledge and knowhow that they had acquired at other places. Via their personal and collegial relationships, they bring to the company, at a very low cost, additional and important knowledge relevant to the Core's goal, which in this case acts as a centripetal (attractive) force with its own "magnetic" field, so to speak, which allows the "funneling" of knowledge into the Core. This knowledge can have a profound impact in the research and save time, human resources, and money. In some cases, as we saw in Chapter 2, it can bring money too!

Therefore, innovation, as demonstrated in this case study, has to meet competent and adequate leadership to bear fruit. Moreover, as mentioned earlier, the Core on its own cannot accomplish everything; it needs external input and a Bridge to access other sources of crucial knowledge and resources. This shall be the subject of the next chapter.

Endnotes

1. Bower, J., Christensen, C., 1995. Disruptive Technologies: Catching the Wave. Harvard Business Review, January–February.
2. Chandler, A. D., 2005. Shaping the Industrial Century: The Remarkable Story of the Evolution of the Modern Chemical and Pharmaceutical Industries. Harvard University Press.
3. Vagelos, R., Galambos, L., 2004. Medicine, Science, and Merck. Cambridge University Press.
4. DiMasi, J.A., Grabowski, H.G., Hansen, R.W. 2016. Innovation in the pharmaceutical industry: new estimates of R&D costs. Journal of Health Economics 47 (5):20–33.
5. PhRMA. 2016. 2016 Biopharmaceutical Research Industry Profile. Washington DC: Pharmaceutical Research and Manufacturers of America. http://phrma.org/sites/default/files/pdf/biopharmaceutical-industry-profile.pdf//
6. Wong, C.H., Siah, K.W., Lo, A.W, January 31, 2018. Estimation of clinical trial success rates and related parameters. Biostatistics. https://academic.oup.com/biostatistics/advance-article/doi/10.1093/biostatistics/kxx069/4817524; see also, https://labiotech.eu/medical/clinical-trials-success-rate/.

7. Booth, B., Zemmel, R., May, 2005. Prospects for productivity. Nature Reviews Drug Discovery 3, 451–456.
8. Hopkins, A., Groom, C., 2002. The druggable genome. Nature Reviews Drug Discovery 1, 727–730.
9. Sánchez-Serrano, I., 2011. The World's Health Care Crisis: From the Laboratory Bench to the Patient's Bedside. Elsevier, 2011.
10. Hu, Y., Thomas Scherngell, T., Man, S.N., Wang, Y., 2013 Is the United States still dominant in the global pharmaceutical innovation network? PLoS one 8(11), e77247. https://doi.org/10.1371/journal.pone.0077247.
11. Lyman, S., September 2, 2014. Which Countries Excel in Creating New Drugs? It's Complicated. Xconomy. https://xconomy.com/seattle/2014/09/02/which-countries-excel-in-creating-new-drugs-its-complicated/.
12. Engelberg, A., October 29, 2015. How Government Policy Promotes High Drug Prices. Health Affairs, https://www.healthaffairs.org/do/10.1377/hblog20151029.051488/full//
13. Hartman, M. et al. 2018. National health care spending in 2016: spending and enrollment growth slow after initial coverage expansions. Health Affairs 37 (1), 150–160; https://www.healthaffairs.org/doi/pdf/10.1377/hlthaff.2017.1299.
14. Goldman, D., Lakdawalla, D., January 30, 2018. The global burden of medical innovation. Report, USC-Brookings Schaeffer Initiative for Health Policy, https://www.brookings.edu/research/the-global-burden-of-medical-innovation/.
15. https://www.merriam-webster.com/dictionary/core.
16. Oliver, H., 2011. Thinking about the firm: a review of Daniel Spulber's the theory of the firm. Journal of Economic Literature. 49 (1): 101–113.
17. Bolton, Patrick; Dewatripont, Matthias 2005. Contract theory. MIT Press.
18. Singer, C., Underwood, A., 1962. A Short History of Medicine. Clarendon Press. See also, Weatherall, M., 1990. In Search of a Cure: A History of Pharmaceutical Discovery. Oxford University Press, New York.
19. Gardner, W.M., 1915. The British Coal-Tar Industry: Its Origin, Development, and Decline. William and Norgate, London.
20. Chapter 3 in Sánchez-Serrano, I., The World's Health Care Crisis.
21. Dupont, A.D. Ibid.
22. Innovation can also be born in a large pharmaceutical company and developed internally or licensed out to external parties.

CHAPTER 4

"The Bridge": collaboration is the best catalyst to success

Introduction

I will start this chapter with a hypothetical and mundane analogy.

Let us imagine for a moment that I live in Boston and that I am interested in meeting a specific philanthropist in San Francisco to present him my business proposal and "pitch" for an early stage biotech start-up to treat a very rare disease that in fact runs in his family. Let us assume that I am interested in securing his support to start the company, cover the licensing fees from a given academic institution, and carry out the first preclinical studies. The reason for this strategy is that most venture capital (VC) investors in the Boston area would not be interested in investing at this early stage, so we need to find alternative sources of funding to create value and attract VC interest. Moreover, the disease we are targeting is so rare that it affects only one 1:250,000 people in the world. Imagine that I have tried to reach out to his office, but to no avail. However, I am told that my prospective funder will be present at a public event to which I can have access free. I have the idea of taking a flight to San Francisco and trying to meet and talk to that person at that event, even if briefly, about my project. However, my resources are very limited and I risk spending my time and money (with food, plane ticket, hotel, and transportation) and return to Boston with nothing, because there is no guarantee that I will be able to speak with him at all.

A few hours later, after these ideas were going through my head, I meet my friend Wendy for coffee. Wendy works for a health care foundation in Boston. I tell her about my project and my interest in meeting the aforementioned philanthropist and my concerns about traveling to San Francisco. Surprised, Wendy asks me why I had not told her before, because back in college she was the roommate of this philanthropist's

The Core Model
ISBN 978-0-12-814293-6
https://doi.org/10.1016/B978-0-12-814293-6.00004-4

personal assistant! Wendy promised that the following morning she would give Lisa, her former roommate, a call on my behalf.

The following afternoon, Lisa calls me to ask me the reason why I would like to speak in person with her boss. I tell Lisa that my *goal* is to discuss with the potential benefactor my project and to explore the possibility of securing funding from his foundation for the further development of an orphan-disease drug that we are interested in targeting and which, by the way, affects his family. Lisa asks me to send some written material and a letter explaining what I had just told her and emphasized that although she could not promise anything she would talk to her boss the following day, before he departs for a trip to India. Three days later I receive a call from Lisa telling me that the philanthropist was interested in meeting with me the following week, upon his return from India. And so we do. Eventually, the philanthropist's Foundation decided to finance my project.

Therefore, because of my coffee with Wendy (someone who had some knowledge of and sympathy for my work on health care, because she herself works in the field and is interested in orphan diseases) I was able to succeed. She could not help me directly, but she knew someone who could probably help. Finally, through these "bridges," or intermediaries, I was able to reach my goal of meeting the philanthropist and secure funding for my "Core."

Needless to say that, without Wendy, I was headed in the wrong direction and it would have taken me a long time (and my own money) to reach out to Mr. Philanthropist—if at all! It would have taken a long time for him, too, if at all, to know me and my enterprise. In the end, not only did I benefit from my connection with Wendy, but she eventually benefited from my work, as she learned more about orphan-disease drugs. Lisa was very grateful to Wendy for introducing me, because her boss was very with happy with the introduction. No money was involved in any of these transactions with Wendy and Lisa. In fact, Mr. Philanthropist was so generous that he reimbursed me for all my traveling expenses!

Although this story is fictitious, this is how things work in real life. We are never independent. We are interdependent and the more personal relationships that are positive we have, the better. This is true with family and friends as it is true professionally. It is important never to underestimate a "connection," no matter how small or insignificant it may seem to be. Down the road, that person can make wonders. Not in vain goes the old adage: "Never burn bridges!" In the previous chapters, we described how Myogenics/ProScript capitalized on its personnel and their connections to secure knowledge relevant for the development of their innovation—a first-in-class anticancer drug candidate that inhibited the proteasome—against all odds. Here, we will analyze how these connections had an impact in the Myogenics/ProScript Core and how this can be applied by other firms.

The role of the Bridge within the Core Model is very similar to the one of an enzyme or catalyst: collaboration increases the speed of a reaction (the outcome) by lowering the activation energy (total amount of effort, time, and investment) without getting consumed in the process (see Figs. 4.1 and 4.2). In this particular case, Myogenics/ProScript went from the creation of inhibitors to block the proteasome ("High-Energy State") to FDA approval of bortezomib for the treatment of multiple myeloma ("Low-Energy State"). One can define a "High-Energy State" within the context of drug discovery and development as the stage in which a product has not yet

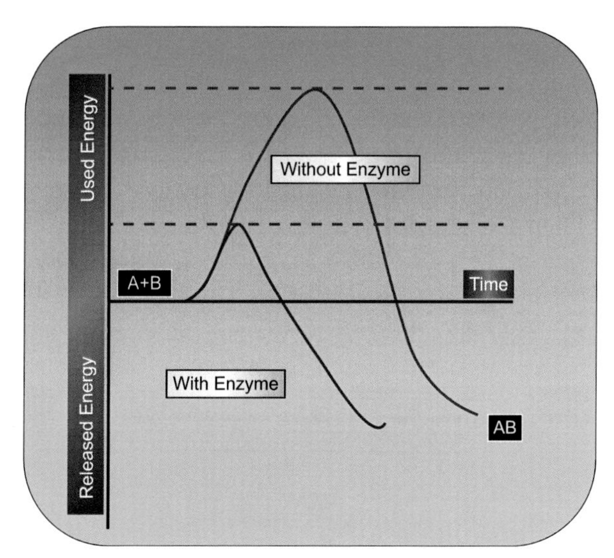

Figure 4.1 ***Mechanism of the Bridge.*** The Bridge represents the immediate collaborators and the private institutions to which the Core has indirect access through the external collaborators. The Bridge contains external scientists interested in similar problems or whose research would be greatly enriched because of the collaboration. It also includes consultants and Scientific Advisory Board (SAB) members (non founders) working in exclusive and non exclusive ways. The mechanism by which the Bridge works is very similar to an enzyme; that is, it acts as a catalyst lowering the activation energy and bringing people together in situations in which otherwise they would not come into contact.

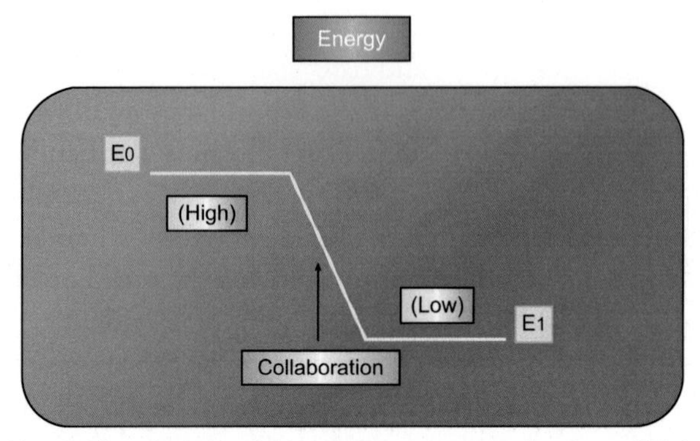

Figure 4.2 ***The Bridge lowers the activation energy from a "High-State" to a "Low-State."*** The role of the Bridge (collaboration) within the Core Model is very similar to the one of an enzyme or catalyst: collaboration increases the speed of a reaction (the outcome) by lowering the activation energy (total amount of effort, time, and investment) without getting consumed in the process.

reached "proof-of-principle" in humans, therefore the need for an enormous amount of effort (labor), time, and investment. So, collaboration (equivalent to an enzyme) lowers the "Energy" required for further progress (proof of principle in animal models and humans, approval, sales, marketing, and profits). The Bridge, as expected by extrapolation, can also be very valuable in other stages, for instance, when going from Phase II (A "High-Energy" State) to Phase III ("Low-Energy" State) and so on.

Albeit the name and the concept of the "Bridge" within the context of the biopharmaceutical industry, as we are discussing in this book, are novel; in practice the Bridge has existed not only in the industry from its very beginning[1] but also from the moment man began living in society. Barter (exchange of goods or services for other goods or services without using money) and trade (the business of buying and selling or bartering commodities or dealings between persons and groups), as defined in economics, are clear examples of this. Without the Core and the Bridge, economically based society (the Periphery) would not exist, as we shall see later.

Academia—industry relationship

Nowhere in the life sciences and biopharmaceutical world is the Bridge more important and relevant than in the academia—industry relationship.

The way in which academia and industry interact is a collaborative and symbiotic process, as I have described in *The World's Health Care Crisis*. Although the traditional goals, views, and interests of each of these sectors are different—for instance, academia is mainly focused on basic research, though in some cases it pursues applications as well; whereas industry does the opposite—in reality they complement each other.

Again, the story case of the development of bortezomib is a perfect example of how the academia–industry relationship can work in an optimal manner. It demonstrates that knowledge and technologies born in academia out of basic research (funded by public and private sources) can continue to have a productive relationship with their parental academic institution as well as with other academic centers, through a feedback-loop (bidirectional) mechanism in which both sides, science and society, are all benefited. However, it is necessary to say that the nature of the collaboration agreements between academia–industry (or academia–academia or industry–industry) can be open (that is, cooperative) or closed (competitive).

One could illustrate a set of relationships between academia and industry through the following Institutional Regime Quadrant below[2]:

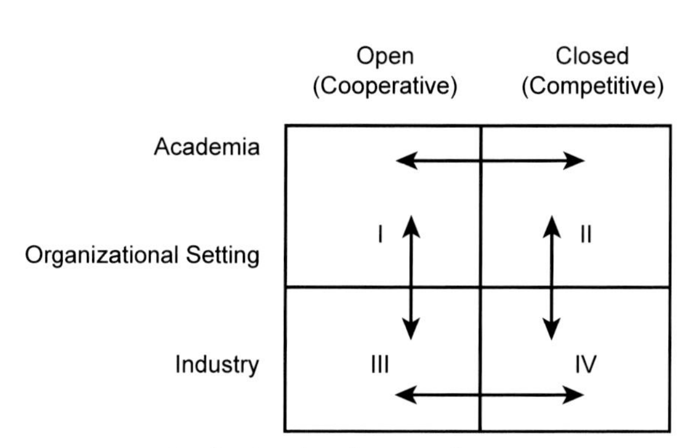

Institutional regime quadrants.

The **first Quadrant** is an open, public, and cooperative system that involves the work of academic scientists doing traditional research, cooperating (maybe) across labs on interesting problems. This kind of work belongs to the public domain; it is sponsored research but with publications generated; it is presented at conferences across universities and it generates international discussions. The **second Quadrant** is a closed, private, and competitive system in which academia-derived companies work with academic scientists in exclusive, secret, or protected relationships via patenting, secrecy, exclusive consulting, and exclusive Scientific Advisory Board memberships. The **third Quadrant** is also an open, public, and cooperative work between firms in industry to bring results for the benefit of the public domain. The **fourth Quadrant** is a private, cooperative system in which firms work together in "private" relationships subject to contracts, patents, New Drug Applications (NDAs), secrecy, hiring, etc.

This type of arrangement is important, because it demonstrates that advancement in drug discovery and development and scientific progress in general can be achieved whether the collaboration between the Core and the Bridge is "open" or "closed" or whether it is "cooperative" and "competitive." Understanding this allows Cores to decide which way they want to go: for profit or not for profit; closed innovation or open innovation. It is also important to know how to identify the kind of collaborators that one will select, as the end goal is to establish a win-win situation based on the "gray zone" of mutual interests among players (see Fig. 4.3)

Tables 2.1 and 4.1 exemplify how some key players in the Bridge contributed to the development of bortezomib. The purpose of these charts is to illustrate that even early on in the process of creating a biotech firm to develop a drug or during strategic alliances and collaboration in big pharma, one can actually create a matrix of the multiple people one would like to collaborate with the specific assets can be exchanged or acquired as to benefit the outcome and save time and money. This can also be done using sophisticated software matching publications, molecules, academic labs, and other information in the public domain. Furthermore, this table demonstrates that without the Bridge the development of bortezomib—and the discovery that it was efficacious in multiple myeloma—would have been impossible. The Bridge saved Myogenics/ProScript an enormous amount of time, brought knowledge and expertise at a very low cost, lowered the activation energy, and even brought funding. Table 4.2 summarizes the benefits of the Bridge. Box 4.1 provides a historical perspective of the importance of the Bridge in the beginning of the pharmaceutical and biotechnology industries.

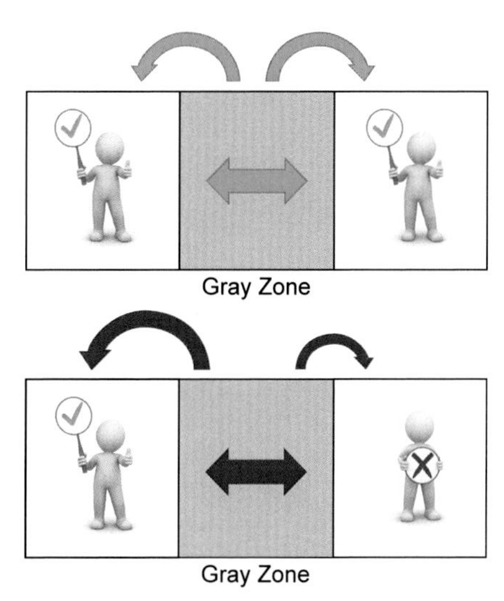

Figure 4.3 ***Dynamics of collaboration.*** When embarking in collaboration, it is important to know how to identify the kind of collaborators that one will select, as the end goal is to establish a win-win situation (top diagram) based on the "gray zone" of mutual interests among players. Being greedy, dishonest, and selfish (lower diagram) creates a poor dynamic that blocks collaboration and creates a poor reputation—and this has a negative impact for society.

Other types of Bridges

No company today has all knowledge and technologies necessary for the understanding of human diseases and for the discovery and development of a given drug. Companies are interdependent with academia, other companies, regulators, federal funding agencies, and other sectors of society, as we saw in the previous chapters, and they rely on the training, expertise, and the connections of their scientists for knowledge generation. In real-world biopharma, one could encounter other types of bridges, such as Contract Research Organizations (CROs), specialized biomarker consortia, tissue banks, Contract Medical Organizations (CMOs), pharmaceutical companies, academic laboratories, investors, advocacy groups, external consultants, incubators, catapults, etc. Any person or entity that can lead to the acquisition of knowledge can be a Bridge.

Table 4.1 Selection of assets exchanged in the Bridge in the development of bortezomib.

Bridge	Asset	Contribution
External collaborators in academia given MG-132 inhibitors	Cell-based assays, animal models	Allowed measurements of the effect of MG-132 in vitro and in vivo, which was crucial in dosing schedules and safety profiles in pre-clinical animal models, as well as information on inhibitor—target association
Alfred Goldberg Lab	Proteasome	Test of the effect of the inhibitors in the proteasome
Kenneth Rock's Lab	Animal models in immunology	Effects of the proteasome in immunology
Tom Maniatis' Lab	Studies on NF-κB	Information on the involvement of the proteasome in the activation of NF-κB, and, in turn, inflammation
Beverly Teicher, DFCI (via SAB member, Bruce Zetter)	Mouse models in cancer	Proof of concept on cancer in the mouse model and the proof that bortezomib worked against tumor growth and metastasis in the mouse model
Avram Hershko (introduced by Julian Adams)	World-leading expert in cancer	Suggested to Julian Adams exploration of cancer as a new business model
David Livingston (introduced by Tom Maniatis)	World-leading expert in cancer	Advice on cancer research. Introduced the company to Kenneth Anderson, a leading figure in cancer at the Dana—Farber Cancer Institute, and who would eventually lead clinical trials.
Hoechst Marion Roussel	Large pharmaceutical company	Agreement reached for up to US$38 million from Hoechst Marion Roussel, plus royalties paid to ProScript on sales of products deriving from this partnership

Table 4.1 Selection of assets exchanged in the Bridge in the development of bortezomib.—cont'd

Bridge	Asset	Contribution
Hoffmann–La Roche	Large pharmaceutical company	US$20 million in equity investment from Roche Group to ProScript, plus royalties on sales of products resulting from this collaboration
Edward Sausville	Head of "Developmental Therapeutics Program" (DTP), National Cancer Institute (NCI)	Access to NCI somatic cell hybrid panel. Funding for Clinical Trials Phase I
Christopher Logothetis at MD Anderson (Introduced by David McConkey, SAB) Important for clinical trials in cancer	Oncologist at MD Anderson cancer center	Conducted Clinical Trials that permitted detection of efficacy in multiple myeloma in Phase I Clinical Trials
Howard Soule (Introduced by Christopher Logothetis)	Chief of Science Officer at CaP Cure (Prostate Cancer Foundation)	Funding for Phase I Clinical Trials in Prostate cancer
David Spriggs	Oncologist at Memorial Sloan Cancer Center	Funding and execution of Phase I Clinical Trial in cancer
Robert Orlowsky	Hematologist at the University of North Carolina, Chapel Hill	Funding and execution of Phase I Clinical Trial; it was this trial that led to the discovery of the efficacy of bortezomib in multiple myeloma in a patient
Kathy Giusti (Introduced by Kenneth Anderson)	Founder of Multiple Myeloma Foundation	Funding and advocacy
Susan Novis (Introduced by Kenneth Anderson)	Founder International Myeloma Foundation	Funding and advocacy

Table 4.2 Summary of the benefits of the Bridge.

Bridge	Benefits to the Core
External collaborators in academia or industry working in exclusive and non-exclusive manner	Personal connections (academia, biotech, pharma, hospitals, regulators, investors, philanthropic organizations, etc.), tests and animal models, clinical trials and funding for clinical trials, tissue banks, new/complementary technologies, reagents, new approaches, and paths to drug discovery and development, new perspectives to improvement, "free" or low-cost labor force (via collaboration), scientific knowledge, problem solving, technological facilities (e.g., mice facilities), special software, cell lines, advice, translational studies. All of this leads to savings time, money, and effort. It increases knowledge and data that most of the time lead to important publications.

In a broader sense, even technologies such as Internet, mass media, social networks, or a scientific journal can become a Bridge. For instance, thanks to the Internet, we can have expedited access to so much knowledge that, without the Internet, would take years and an extraordinary amount of labor and capital to access. Similarly, thanks to mass media (Radio, TV, newspapers, books, and magazines), we can know what happens at any single point of the world without traveling and, in some occasions, as in the case of abuse or oppression, we can make our voices heard in society and injustices known, so a remedy can be found. Thanks to scientific journals, we can find out information about chemical, physical, and biological processes elucidated by others, without having to do the work. Importantly, it is through the Bridge that we can have access to the Periphery, that is, all the resources created by society for its own well-being, which we shall discuss in the following chapter.

BOX 4.1 Historical perspective on the use of the Bridge

From its very beginnings in the mid-19th century in Europe and the United States, the pharmaceutical industry has relied for its success on the close interaction and fluid cooperation between academia and industry and the public and private sectors. This was epitomized by legendary companies such as Hoechst, Bayer, Merck & Co., Afga, Ciba, Sandoz, Parke–Davis, Hoffmann–La Roche, and Burroughs Wellcome, among many others[3,4,5,6]. The collaboration between these sectors—despite the secrecy, Intellectual Property Rights, and economic interests intrinsic to the drug industry—has resulted in immeasurable benefits for science and society, through the generation of basic scientific knowledge and the creation of very important drugs that have revolutionized medicine and health care and changed the history of humankind.

Collaboration with external players and outside academic groups has been essential to all biopharmaceutical companies from the very beginning of this industry. For instance, the success of the German chemical and pharmaceutical company Hoechst lies in its ability to secure strong ties with academia and academic scientists—such as Robert Koch (and later with Clemens von Pirquet and Koch's student Arnold von Libbertz) for the development of tuberculin as a testing agent for tuberculosis. Others include Emil von Behring, who discovered the diphtheria antitoxin and developed a serum therapy against diphtheria (together with Emile Roux) and tetanus; and, among many others, Paul Ehrlich, who postulated the theory of receptors, coined the term chemotherapy, found a cure for syphilis (Salvarsan), and carried out impressive studies on autoimmunity. The development of these agents in a commercial manner was the direct result of close collaboration between the "Core" (in this case academic inventors and Hoechst), the "Bridge" (other academic scientist working in related problems), and the "Periphery" (public hospitals, the German government, and other not-for-profit organizations). Other German firms (such as Merck & Co., Afga, and Bayer), the major Swiss firms (Ciba, Hoffmann–La Roche, Sandoz), and European and American firms followed a similar pattern in the second half of the 19th century, cementing the great success of the pharmaceutical industry ever since. In the twentieth century, medicines such as penicillin (first produced in a mass scale by Merck & Co. in the United States), insulin (commercially released by Eli Lilly), the first sulfonamide, Prontosil (Bayer), and antibiotics, in general, were also developed in the same fashion. Late in the 20th century, the birth of the biotechnology industry and the development of the first biotechnology drugs can be explained by the Core Model, with a close and systematic collaboration between academia, industry, investors, government agencies, and philanthropic organizations/advocacy groups, via knowledge transfer, integration, and translation.

Endnotes

1. For a comprehensive story of the chemical and pharmaceutical industry please, refer to Chandler, A.D., 2005. Shaping the Industrial Century: The Remarkable Story of the Evolution of the Modern Chemical and Pharmaceutical Industries. Harvard University Press.
2. This framework was inspired by Donald, E. Stokes' Pasteur's Quadrant. Brookings Institution Press, Washington, D.C., c1997.
3. Sánchez-Serrano, I. 2016. The World's Health Care Crisis: From the Laboratory Bench to the Patient's Bedside. Elsevier.
4. Weatherall, M. 1991. In Search of a Cure: A History of Pharmaceutical Discovery. Oxford University Press.
5. Chandler, A.D. 2005 Shaping the Industrial Century: The Remarkable Story of the Evolution of the Modern Chemical and Pharmaceutical Industries (Harvard Studies in Business History). Harvard University Press.
6. Raviña, E. 2011. The Evolution of Drug Discovery: From Traditional Medicines to Modern Drugs. Wiley-VCH.

CHAPTER 5

"The Periphery": the efficient use of underused societal resources

Since its beginning, human society has created a significant variety of mechanisms, institutions, and resources to safeguard the well-being of its members. Over the millennia, these mechanisms, institutions, and resources have evolved into what we now know as public institutions such as governments, schools, universities, hospitals, research centers, foundations, regulatory agencies, public libraries, philanthropic organizations, museums, parks, churches, etc., and a system of rules enforced through social institutions to govern behavior, namely laws—for the benefit of all its citizens. Today, most of these entities—regardless of the type of government—are financed via the tax contribution of the members of society or through the wealth generated by all. Even philanthropy is a "devolution" made by some very conscientious or altruistic individuals to society for all the benefits they have received. Without these institutions and mechanisms, which I have termed the "Periphery," meaning "surroundings" or outer limits of the Core and the Bridge, society would not exist. In fact, the Periphery represents the very interests of society.

We have already discussed that the process of developing novel drugs to cure and treat diseases requires remarkable innovation and leadership. We have also highlighted that drug discovery and development are very labor-, time-, and capital-consuming activities. Moreover, they are highly regulated as well. Probably no other industry in the world is as highly regulated as the biopharmaceutical industry, and for obvious reasons. All these factors, in fact, pose great limitations on whoever is willing to take the high risk of discovering and developing new drugs, as they are natural barriers or deterrents to entry to many entrepreneurs or even smaller companies. For a start-up wishing to begin drug discovery and development, from the time of company incorporation to the point of selecting a drug candidate in the preclinical studies, it may take no less than US$ 5–10 million and 2–3 years and a significant amount of work, planning, and coordination. However, it is important to emphasize that the Periphery

The Core Model
ISBN 978-0-12-814293-6
https://doi.org/10.1016/B978-0-12-814293-6.00005-6

encompasses an important number of resources that help mitigate the risk, time, labor, and expense of drug discovery and development for smaller companies as well as for larger ones. Examples include universities and teaching institutions, research centers, hospitals, drug regulators, philanthropies and charities, consortia, federal funding agencies, legal systems protecting intellectual property, financial and fiscal institutions and regulators, private equity, intellectual property laws, ethical regulations, lobby groups, etc. Sometimes these resources are well utilized, but often they are underused, underfunded, unknown, or misunderstood. Though it would seem easy to assume that any given individual could have ready access to the resources of the Periphery, in reality and in many circumstances, this can be rather difficult, due to large demand and the limited variety of resources available. In most instances, as we have already seen, access to Periphery resources occurs through the Bridge: intermediary individuals, collaborators, advocates, and even intermediary electronic platforms.

In the case of the development of bortezomib, we saw how Myogenics/ProScript was able to secure resources from the Periphery (universities, funding institutions, government agencies, hospitals, foundations, etc.) through the Bridge in the most efficient way. This allowed them to not only acquire essential knowledge but also be able to capitalize on public resources and funding, such as clinical trials sponsored by University of North Carolina and Memorial Sloan Kettering. Others included the somatic cell hybrid panel provided by the National Cancer Institute (NCI), and funding by the Prostate Cancer Foundation (CaP Cure), the Multiple Myeloma Foundation, the International Multiple Myeloma Fund, and social awareness by such institutions, etc. This is a clear demonstration how a small company can avoid depleting its limited resources via these strategic alliances or collaborations. Even Big Pharma can benefit in funding and collaborating in the creation of large platforms, such as open-source tissue banks, biomarkers, consortia, databases, and programs, such as the human genome project or resources to facilitate the execution of clinical trials. In addition, translational research, and even social programs related to health care education and affordable access to medicines, etc., could have a major impact on society in terms of facilitating drug research and approval of novel compounds, and eventually in increasing access to novel medications by the world population.

The Periphery is sustained thanks to public funding (tax contribution, wealth redistribution, and international aid and cooperation), private funding (philanthropy, corporate social responsibility), or a mixture of the two. In the United States, some of the most important constituents in the Periphery related to our analysis in this book include: federal funding agencies (such as the National Institutes of Health [NIH] and the National Science Foundation [NSF]), regulatory bodies (such as the FDA), philanthropic organizations (instance.g., disease-related foundations), biomarker consortia, Big Pharma resources, and advocacy groups. Let us discuss some of them in further detail and explore how they could contribute to the creation of new drugs. Due to space limitations, we will just discuss the most relevant ones, as it is impossible to list them all in a book chapter, and, here, we are mostly interested in the concept.

Publicly funded Periphery

The National Institutes of Health (NIH)[1]

Founded in 1870, the NIH is a division of the U.S. Department of Health and Human Services. As the US Government's agency responsible for biomedical and public health research, NIH plays an essential role in fostering innovation by funding mostly purely basic research and some translational research through the NIH Roadmap for Medical Research. This is an initiative launched in 2004 tasked with transforming the way biomedical research is conducted in the United States. The Roadmap fosters high-risk/high-reward research, closing the knowledge gaps in some specific and crucial areas, encouraging academic collaboration with outside partners, and developing novel tools and methodologies to tackle research problems that no single NIH institute can address alone. Funded through the NIH Common Fund, the programs included in the Roadmap initiative cover all areas of health and disease research within the NIH.

This US government agency is the largest governmental funding body in the world. For Fiscal Year 2018 (FY 2018) the NIH budget was US$ 37 billion, its largest in 15 years.[2] As the NIH webpage describes, more than 83% of the NIH's funding is awarded through almost 50,000 competitive grants to more than 325,000 researchers at more than 3000 universities, medical schools, and other research institutions in every state and around the world. About 10% of the NIH's budget supports projects conducted by nearly 6000 scientists in its own laboratories, most of which are on the NIH campus in Bethesda and Rockville, Maryland. The NIH comprises 27 separate institutes, of which the first and largest is the National Cancer Institute, established in 1937, and six centers. The latest of these is the National Center for Advancing Translational Sciences (NCATS) the mission for which is to catalyze the generation of innovative methods and technologies that will enhance the development, testing, and implementation of diagnostics and therapeutics across a wide range of human diseases and conditions,[3] was established as recently as 2011. Over many decades, the NIH has been a valuable resource for companies in their quest for new drugs by being very open at maintaining both formal and informal relationships with pharmaceutical companies and early-stage biotechnology companies. Companies with connections to the NIH often gain rights to agents and drug targets discovered by the NIH (usually in conjunction with a leading university). For example, Bristol–Myers Squibb's Taxol anticancer drug originated from NIH research efforts.

Some of the recent and ongoing initiatives by the NIH to accelerate drug discovery and development include the Cancer Moonshot, the Accelerating New Medicines Partnership, the Brain Research through Advancing Innovative Neurotechnologies, and the All of Us Research Program, among several others. Let us briefly describe them, although information that is more detailed can be found at the NIH website.

- **Cancer Moonshot:** This initiative was passed by the U.S. Congress as the 21st Century Cures Act in December 2016. This act authorizes the allocation of US$1.8 billion in funding for the Cancer Moonshot over 7 years. Its goal is to accelerate cancer research and make therapies available so that more patients benefit from novel cancer treatments. Cancer prevention and early detection also form an integral part of this initiative. The funding must be appropriated each fiscal year over those 7 years. Congress appropriated $300 million to NCI for fiscal year (FY) 2017, $300 million for FY 2018, and $400 million for FY 2019. See Table 5.1 for a description of the initiatives approved by the Cancer Moonshot Blue Ribbon Committee.
- **Accelerating New Medicines Partnership (AMP):** This initiative was launched in 2014 as a public–private partnership between the National Institutes of Health (NIH), the U.S. Food and Drug Administration (FDA), 12 biopharmaceutical and life science companies, and 13 nonprofit organizations. The objective of the AMP is the transformation of the current model for developing new treatments and diagnostics in three major disease areas: Alzheimer's disease (AD), type 2 diabetes (T2D), and autoimmune disorders (i.e., rheumatoid arthritis and systemic lupus erythematosus). This joint approach pursues the identification and validation of promising biological targets for therapeutics and in this way develop these technologies for the benefit of

Table 5.1 Cancer moonshot research initiatives.

Following receipt of the Blue Ribbon Panel (BRP) report, and the authorization of the 21st century Cures Act, the NCI established implementation teams that align with each of the BRP recommendations. The teams have identified opportunities and developed initiatives for funding that directly address each of the recommendations. These mark the beginning of a Cancer Moonshot portfolio that will continue to be expanded in future years.

The following initiatives have been established to address the goals of the recommendations:

- Establish a Network for Direct Patient Engagement
  Engage patients to contribute their comprehensive tumor profile data to expand knowledge about what therapies work, on whom, and on which types of cancer.
- Create an Adult Immunotherapy Network
  Establish a cancer immunotherapy research network to develop immunity-based approaches for the treatment and prevention of cancer in adult patients.
- Create a Pediatric Immunotherapy Discovery and Development Network (PI-DDN)
  Generate a cancer immunotherapy research network to overcome challenges in the development of immunotherapies for childhood cancers.
- Develop Ways to Overcome Cancer's Resistance to Therapy
  Identify therapeutic targets to overcome drug resistance through studies that determine the mechanisms that lead cancer cells to become resistant to previously effective treatments.
- Build a National Cancer Data Ecosystem
  Create a national ecosystem for sharing and analyzing cancer data so that researchers, clinicians, and patients will be able to contribute data that will facilitate efficient data analysis.
- Intensify Research on the Major Drivers of Childhood Cancers
  Improve our understanding of fusion oncoproteins in pediatric cancer and use new preclinical models to develop inhibitors that target them.
- Minimize Cancer Treatment's Debilitating Side Effects
  Accelerate the development of guidelines for routine monitoring and management of patient-reported symptoms to minimize debilitating side effects of cancer and its treatment.
- Prevention and Early Detection of Hereditary Cancers
  Improve current methods and develop new strategies for the prevention and early detection of cancer in individuals at high risk for cancer.
- Expand Use of Proven Cancer Prevention and Early Detection Strategies
  Reduce cancer risk and cancer health disparities through the development, implementation, and broad adoption of proven cancer prevention and detection approaches.
- Analyze Patient Data and Biospecimens from Past Clinical Trials to Predict Future Patient Outcomes

Continued

Table 5.1 Cancer moonshot research initiatives.—cont'd

Predict response to standard treatments through retrospective analysis of patient specimens.

- Generation Human Tumor Atlases
  Create dynamic 3D maps of human tumor evolution to document the genetic lesions and cellular interactions of each tumor as it evolves from a precancerous lesion to advanced cancer.
- Develop New Cancer Technologies
  Develop new enabling cancer technologies to characterize tumors and test therapies.

Partnership for Accelerating Cancer Therapies (PACT)
PACT is a 5-y, public–private research collaboration launched by NIH and 11 pharmaceutical companies. As part of this effort, NCI recently awarded grants to support four Cancer Immune monitoring and Analysis Centers (CIMACs) and a Cancer immunologic Data Commons (CIDC).

From National Cancer Institute (NCI); https://www.cancer.gov/research/key-initiatives/moonshot-cancer-initiative/implementation *(original source).*

patients, while aiming simultaneously at reducing the time and cost of development. An AMP project on Parkinson's disease (PD) was launched with nine partners in January 2018.
This partnership is managed through the Foundation for the NIH (FNIH). Industrial partners and the NIH are sharing expertise and over $350 million in resources (which include in-kind contributions). All partners in this joint effort have agreed to make the AMP data and analyses publicly accessible to the broad biomedical community, which is a very valuable asset. (See Tables 5.2 and 5.3 for more details on AMP.)

- **The Brain Research through Advancing Innovative Neurotechnology (BRAIN) Initiative:** This project has as a goal to gain knowledge on how the human brain functions. Through the acceleration of the development and application of innovative technologies, scientists will be able to see for the first time, and in real time, how individual cells and complex neural circuits interact in both time and space. The objective of this approach is to create a "dynamic picture" of the brain, bridge major gaps in our understanding of how this organ works, and explore in a more detailed way how the brain enables the human body to record, process, utilize, store, and retrieve vast quantities of information, all at the speed of thought. This initiative will have important implications in the treatment, cure, and even prevention of

Table 5.2 List of AMP partners as of 2018.

Government	Industry	Non profit organizations
FDA NIH	AbbVie Biogen Bristol–Myers Squibb Celgene GlaxoSmithKline Johnson & Johnson Lilly Merck Pfizer Sanofi Takeda Verily	Alzheimer's Association Alzheimer's Drug Discovery Foundation American Diabetes Association Arthritis Foundation Foundation for the NIH Geoffrey Beene Foundation Juvenile Diabetes Research Foundation Lupus Foundation of America Lupus Research Alliance Michael J. Fox Foundation for Parkinson's Research PhRMA Rheumatology Research Foundation US Against Alzheimer's

From National Institutes of Health (NIH); https://www.nih.gov/research-training/accelerating-medicines-partnership-amp#partners *(original source).*

Table 5.3 Current AMP funding commitments (total: 5 years).[a]

Disease area	Total NIH funding ($M)	Total industry funding ($M)	Total nonprofit funding ($M)	Total project funding ($M)
Alzheimer's Disease (AD)	162	22.2 (+40 in kind)	1.0	185.2 (+40 in kind)
Type 2 Diabetes (T2D)	31	21.5 (+6.5 in kind)	0.3	52.8 (+6.5 in kind)
Rheumatoid Arthritis and Lupus (RA/Lupus)	20.9	20.7 (+0.1 in kind)	0.6	42.2 (+0.1 in kind)
Parkinson's disease (PD)	12	8 (+2 in kind)	2.0	22 (+2 in kind)
Total ($M)	225.9	72.4 (+48.6 in kind)	3.9	302.2 (+48.6 in kind)

[a]Start date for AD, T2D, and RA/Lupus was February 2014. Start date for PD was January 2018.
From National Institutes of Health (NIH); https://www.nih.gov/research-training/accelerating-medicines-partnership-amp *(original source).*

devastating diseases such as AD, PD, depression, and traumatic brain injury (TBI). Inspired by the Human Genome Project, BRAIN is a collaborative, public–private research initiative announced by the Obama administration on April 2, 2013.

- **Precision Medicine Initiative (PMI)/*All of US* Research Program:** This project was launched in 2016 with the objective of building a national, large-scale research participant group, called a cohort. The NIH was allocated $130 million in funding for that purpose, whereas the NCI was allocated $70 million to lead efforts in cancer genomics. The *All of Us* Research Program is an important component of the PMI. PMI will bring a new era in medicine in which a great deal of information will be gathered from patients so that researchers, health care providers, and patients work together to develop individualized care.

In addition to these initiatives, each institute of the NIH provides the opportunity for academic and industrial collaborators to have access to their facilities, cell lines, protocols, sophisticated instrumentation, animal models, etc.—all of which contribute in great manner to save time, labor, and money, while increasing knowledge about very specific areas on the function of biological systems.

The National Science Foundation (NSF)

The NSF was established by the National Science Foundation Act of 1950, with the mission of promoting "the progress of science; to advance the national health, prosperity, and welfare; and to secure the national defense."[4] Although the NSF is the only US federal agency with a mandate to support *all* nonmedical fields of research, with an annual budget of $7.8 billion (FY 2018), it funds around 24% of all federally supported basic research conducted by the colleges and universities of the United States. The National Science Foundation is in some cases the major source of funding in disciplines such as mathematics, computer science, and the social sciences. However, all these disciplines, through multidisciplinary collaboration, have a direct and indirect impact on biomedical and public health research.

Although not focused on biomedical research, because its counterpart, the NIH, is responsible for that, the NSF has a Directorate for the Biological Sciences (BIO). The mission for BIO is "to enable discoveries for understanding life" by supporting research that advances the frontiers of biological knowledge, increases our understanding of complex systems, and provides a theoretical basis for original research in many other scientific disciplines. There are five different divisions for biological research at BIO, at present, among which the Division of Biological Infrastructure (DBI) and Division of Molecular and Cellular Biosciences (MCB) are the closest to biomedical research and human disease. DBI supports the development and enhancement of resources for research, human capital, and mid-to-large-scale infrastructure and centers to promote advances in all areas of biological research. MCB, in turn, supports quantitative and interdisciplinary approaches to deciphering the molecular underpinnings of complex living systems.

The Food and Drug Administration (FDA)[5]

The Food and Drug Administration, as we know it, was created in 1938 when, in response to a number of deaths from the use of a poisonous solvent, diethylene glycol, in a new sulfa drug, the U.S. Congress decided that the government should take steps that are more systematic to protect the public. The FDA was assigned the specific task of requiring drug companies to prove that their products were safe before they could be sold. However, Congress delayed until 1952 to decide that a doctor's prescription would be necessary to purchase drugs that could not be used safely without medical expertise. In 1962, it was found that thalidomide (Thalomid), which had been licensed as a tranquilizer in Europe but not in the United States, was responsible for serious birth defects in the offspring of women who had taken the drug during pregnancy. Therefore, another requirement was added to the FDA approval process: drug companies had to prove that their products were not just safe but effective. That mandate soon gave rise to rules for carrying out clinical trials as we know them today. This agency is responsible for protecting the public health by ensuring the safety, efficacy, and security of human and veterinary drugs, biological products, and medical devices; and by ensuring the safety of our nation's food supply, cosmetics, and products that emit radiation. The FDA also has responsibility for regulating the manufacturing, marketing, and distribution of tobacco products and e-cigarettes to protect the public health and to reduce tobacco and e-cigarette use by minors.

FDA is responsible for advancing the public health by helping to speed innovations that make medical products more effective, safer, and more affordable and by helping the public get the accurate, science-based information they need to use medical products and foods to maintain and improve their health. FDA also plays a significant role in the Nation's counterterrorism capability. FDA fulfills this responsibility by ensuring the security of the food supply and by fostering development of medical products to respond to deliberate and naturally emerging public health threats.

Over the years, the FDA has created a series of initiatives to accelerate the drug discovery and development process: For instance:

- **Prescription Drug User Fee Act:** The PDUFA was first enacted in 1992, with the goal of improving the process of drug review. Under this program—which has been subsequently revised in 1997, 2002, and more recently in 2007, as PDUFA II, III, and IV, respectively—the pharmaceutical and biotechnology industries pay certain user's fees to the FDA. The fee of the program for Fiscal Year 2018 was $915,825,714 (However, with other adjustments and carryover monies from the 2017 PDUFA program, the grand total for FY 2018 PDUFA VI FY 2018 fees turns out to be $911,346,000).[6] In exchange for these fees, the FDA agreed, via correspondence with Congress, to a set of performance standards intended to reduce the approval time for NDAs and Biological License Applications (BLAs).
- **Fast Track, Priority Review, and Accelerated Approval:** Sometimes the FDA allows drugs that are still in clinical trials to be administered to seriously ill patients, upon request by the sponsor company, through three different approaches, all with "speed" as a common denominator: fast track, priority review, and accelerated approval. Fast track is a process designed to facilitate development and to expedite the review of drugs to treat serious diseases (such as AIDS, Alzheimer's disease, heart failure, diabetes, depression, and cancer) and to fill an unmet medical need. The purpose is to get important new drugs to the patient earlier. A priority review designation is given to drugs that offer major advances in treatment or that provide a treatment for which no adequate therapy exists. A priority review means that the time it takes the FDA to review a new drug application is reduced. The goal is to complete a priority review within 6 months. Often a company requesting fast track designation for a particular drug can also request priority review. Accelerated approval is a procedure (under the Subpart H regulation) through which the FDA may grant marketing approval for a new drug product on the basis of adequate and well-controlled clinical trials. These clinical trials must establish that the drug product has an effect on a surrogate marker that is reasonably likely to predict clinical benefit, or on the basis of an effect on a clinical end point other than survival or irreversible morbidity. Approval under this section will be subject to the requirement that the applicant study the drug further to verify and describe its clinical benefit, in which uncertainty exists as

to the relation of the surrogate end point to clinical benefit or of the observed ultimate clinical benefit.

- **Orphan-Disease Drugs:** Many companies focus, however, on the development of orphan drugs. In 1983, the FDA approved the Orphan Drugs Act, designed for drugs to treat rare diseases within patient populations of less than 200,000 individuals. This law provides research grants, tax breaks, and exclusive marketing rights for 7 years, and other benefits for companies that develop these kinds of drugs.
- **FDA Critical Path Initiative:** Launched in 2004, the Critical Path Opportunity List Initiative had as a goal of improving and accelerating the process of translating experimental leads into approved medicines and creating significant public health benefits in this manner. A crucial part of the Critical Path involves translational medicine, the bench-to-bedside feedback loop mechanism between basic researchers and clinicians that is considered the bridge between discovery and development. In the translational process, *biomarkers*—that is, quantitative measures that provide the link between mechanism and clinical effect, assisting in the evaluation of targets (i.e., biological pathways in disease causation and prevention) and matching them to investigational compounds—are extremely important.

The FDA, in collaboration with the National Institutes of Health (NIH) and several US-based universities, is particularly focused on developing biomarkers for proteomics, imaging, immune response, liver toxicity, and QT prolongation (i.e., the risk associated with heart ventricular complications). In addition to this, the FDA is open to have conversations with entrepreneurs at very early stages during the development process. Pharmacovigilance and electronic resource of all sorts and a list of open clinical trials, etc. (see Table 5.4 for a list of some of the FDA initiatives).

Privately funded Periphery

In the United States, as well as in most industrialized countries, many private foundations (such as the Bill and Melinda Gates Foundation, the Rockefeller Foundation, The Eli and Edythe Broad Foundation, among many others) pursue a mission of providing funding, expertise, advocacy, education, and platforms to foster drug discovery and development, patient recruitment, patient's resources and support, etc., for a significant number of diseases. The beautiful work that these foundations carry out not only

Table 5.4 Selected list of some of the FDA initiatives to contribute to the acceleration of drug development.

Initiative	Description
Bioinformatics Tools	Tools for analyzing and integrating genomics, transcriptomics, proteomics, and metabolomics datasets.
Health Informatics at FDA	Overview of FDA's health informatics initiatives to optimize acquisition, storage, retrieval, and use of information in public health and biomedicine.
Data Mining at FDA	Information on data mining initiatives at FDA.
INFORMED Program	INternet curriculum FOR Melanoma Early Detection (INFORMED) is an incubator for collaborative regulatory science research focused on supporting innovations that enhance FDA's mission.
Science and Research Special Topics	Women's health, pediatrics, the Critical Path Initiative, nanotechnology, clinical trials, and peer reviews of scientific information.
Funding for Rare Diseases and Pediatric Device Consortia	FDA's role in providing incentives for sponsors to develop products for rare diseases.
FDA Technology Transfer Program	Information to help FDA and potential collaborators transfer federal technology to the commercial marketplace.
Field Science and Laboratories	The Office of Regulatory Affairs (ORA) Laboratory Profiles, Laboratory Manual, Laboratory Information Bulletins, Policy for consumer product tests, training laboratory staff, report writing, private laboratory report reviews, and testimony.
Meetings, Conferences, and Workshops	FDA sciencerelated gatherings, including the Science Forum, Science Writers Symposium, and public science meetings.
Foods and Veterinary Medicine Science and Research	Strategic direction, coordination and methods development for FDA's food and veterinary research.

From Food and Drug Administration (FDA); www.fda.gov *(original source).*

impacts the provision of useful resources for the creation of novel drugs, but they also create interest in diseases that otherwise would not interest the pharmaceutical industry.

As of 2015, there were at least 86,203 patient-assisting foundations in the United States with a total giving amount of almost US$ 63 billion.[7] Some of the largest pharmaceutical companies were among the top 50 giving foundations. Nevertheless, many of these firms also sponsor research in areas other than the ones with mentioned earlier in their collaboration with the NIH, such as tropical medicine (for instance, Novartis, GlaxoSmithKline, Merck, AstraZeneca, etc.) and rare diseases. Others, such as Novartis, Merck, and Eli Lilly, among many others, have venture funds to help start-up companies, and count with or sponsor educational programs of one sort or another, etc. These are all resources that could be useful for drug discovery and development, for the pharmaceutical industry and for global health care.

Public privately funded Periphery

Within this category, we encounter initiatives such as the public–private partnerships. We also have nongovernmental organizations (such as the United Nations Development Program and the World Health Organization) that work with governments and with industry on specific projects related to health care education, prevention, access to medicines, especially for international tropical infectious diseases, crisis management, infrastructure, health care guidelines, information platforms and resources, reports, etc. The Biomarkers Consortium, managed by the FNIH, for instance, is one such public–private biomedical research partnership that seeks to develop and qualify biomarkers across a broad range of diseases to accelerate the development of new medicines and improve patient care.[8,9]

Summary: the Periphery and the Core Model

At present, the Periphery of the Core Model is becoming more dynamic than ever. For instance, patient groups as well as health systems—which are partners of the pharmaceutical industry in clinical trials—are now playing an active role beyond advocacy, including actual financing of drug development, especially in rare diseases. This was the case, for instance, with the Cystic Fibrosis Foundation financing Kalydeco[10] with Vertex, and thus bypassing Big Pharma. Patient opinion leaders and online communities are now collaborating in clinical trials and conducting their own observational studies, as in the case of PatientsLikeMe[11] and its 600,000 members. The importance of patient-reported outcomes in collecting real-world evidence is covered in several policy journals, such as *Health Affairs*[12] and *Value in Health.*[13] These initiatives are becoming a novel and important model for improving drug development.

Frustration with the biopharmaceutical companies' high pricing of drugs, sudden increasing of the pricing of old, off-patent drugs such as the heart medicine Nitropress, and shortages of essential medicines, such as morphine, as well as the manipulation of the market by investors, has led a group of large hospital systems to the bold decision of creating a nonprofit organization. This new nonprofit in the United States,[14] spearheaded by Intermountain Healthcare, and involving Ascension—a Catholic system that is the nation's largest nonprofit hospital group—and very possibly the Department of Veteran Affairs, will create generic drugs to battle shortages and high prices in hospitals. In another bold movement, three large US corporations—Amazon, Berkshire Hathaway, and JPMorgan Chase—have announced the formation of an independent health care company for their employees in the United States to deal with the soaring cost of health insurance premiums and pharmaceuticals.[15]

Two other important initiatives adopting the Core Model are, first, the NIH–National Center for Advancing Translational Sciences' Industry Partnerships Initiative (launched in 2012) to foster collaboration between pharmaceutical companies and the biomedical research community to advance therapeutics development, as part of the NIH New Therapeutics Use Program. The objective of this initiative is to match researchers with a selection of "pharmaceutical assets" to help the scientists test ideas for new therapeutic uses.[16] The second one is Partnership for Accelerating Cancer Therapies (PACT)—a $215 million public–private partnership with 11 pharmaceutical companies over a 5-year agreement to advance research in new immunotherapy treatments that equip the immune system to attack cancer via the identification, development, and validation of biomarkers. This initiative not only involves the partners just mentioned, but also the NIH Foundation, important academic research centers in the United States, the Food and Drug Administration (FDA), Pharmaceutical Research and Manufacturers Association (PhRMA), and the Department of Health and Human Services, among other partners.[17]

All these examples attest to the flexibility of the Core Model and its universality. Today, the biopharmaceutical industry does not know where it is going and faces the extraordinary necessity of coming up with more efficient ways to produce new drugs. Therefore, given the extraordinarily big impact that the high price of novel drugs has in the global health care systems and on some countries' economies, it is extremely important to review the Core Model paradigm and adapt it according to specific circumstances (Box 5.1).

BOX 5.1 Benefits of the Periphery

Funding, infrastructure, promotion, lobbying, recruitment of patients for clinical trials, "free" labor, equipment, positive public image, knowledge, data, know-how, technology, tips by regulators to correct wrong paths, guidance on efficiency, translational science, saving time and money while guiding the Core on how to succeed.

Endnotes

1. See the NIH website for further information: https://www.nih.gov/.
2. Science News Staff, March 23, 2018. Trump, Congress approve largest U.S. research spending increase in a decade. Science Magazine. http://www.sciencemag.org/news/2018/03/final-2018-budget-bill-eases-biomedical-researchers-policy-worries.
3. See NCATS website: https://ncats.nih.gov/.
4. See the National Science Foundation website: https://www.nsf.gov/.
5. See the FDA website: www.fda.gov.
6. Pollock, B., September 13, 2017. PDUFA VI Fee Structure and Fees Announced — Gulp! Lachman Consultants. http://www.lachmanconsultants.com/2017/09/pdufa-vi-fee-structure-and-fees-announced-gulp/.
7. Foundation Center. http://data.foundationcenter.org/#/foundations/all/nationwide/top:giving/list/2015.
8. See Foundation for the NIH-Biomarkers Consortium. https://fnih.org/what-we-do/biomarkers-consortium.
9. Wholley, D., 2014. The Biomarkers Consortium. Nature Reviews Drug Discovery 13, 791—792.
10. Cystic Fibrosis Foundation website: https://www.cff.org/About-Us/About-the-Cystic-Fibrosis-Foundation/CF-Foundation-Venture-Philanthropy-Model/; accessed 02/5/2018.
11. PatientsLikeMe website: https://www.patientslikeme.com/about?utm_source=googlesearch&utm_medium=cpc&utm_campaign=brand_general_search_adgroup_brandedbroad_lp2_h58_h57_copy64&gclid=Cj0KCQjwvuDPBRDnARIsAGhuAmbfGO4gWEDxNW7v6zOd4vQ_79l9yRAVjGZ6somsfLGv8OYOL6pXitAaApaPEALw_wcB; accessed 02/5/2018.
12. Howie, L., Hirsch, B., Locklear, T., Abernethy, A.P., 2014. Assessing the value of patient-generated data to comparative effectiveness research. Health Affairs 33, 1220—1228.
13. Berger, M.L., Lipset, C., Gutteridge, A., Axelsen, K., Subedi, P., Madigan, D., 2015. Optimizing the leveraging of real-world data to improve the development and use of medicines. Value in Health 18, 127—130.
14. Wingfield, N., Thomas, K., Abelson, R., January 30, 2018. Amazon, Berkshire Hathaway and JPMorgan team up to try to disrupt health care. The New York Times. https://www.nytimes.com/2018/01/30/technology/amazon-berkshire-hathaway-jpmorgan-health-care.html.
15. Abelson, R, Thomas, K., January 18, 2018. Fed up with drug companies, hospitals decide to start their own. The New York Times. https://www.nytimes.com/2018/01/18/health/drug-prices-hospitals.html.

16. National Center for Advancing Translational Science. About New Therapeutic Uses. https://ncats.nih.gov/ntu/about.
17. National Institutes of Health (NIH), October 12, 2017. NIH partners with 11 leading biopharmaceutical companies to accelerate the development of new cancer immunotherapy strategies for more patients. Press Release. https://www.nih.gov/news-events/news-releases/nih-partners-11-leading-biopharmaceutical-companies-accelerate-development-new-cancer-immunotherapy-strategies-more-patients.

PART III

CHAPTER 6

Biopharmaceutical intellectual property, financial interests, and the Core Model

Intellectual Property Rights (IPRs) are at the center of the biopharmaceutical industry's business model. Without them—at least within the traditional way by which the pharmaceutical industry operates—investing in, developing, and commercializing innovation would be extremely difficult. Nevertheless, criticisms of the industry have been in place for many years because of its attitudes and practices related to IPRs, which, in many observers' opinions, have gone too far to benefit the industry at the expense of the patients, while blocking innovation by other players.

Specific examples of the pharmaceutical industry's praxis in this respect have recently been provided by Initiative for Medicines, Access and Knowledge (I-MAK), a think tank the mission for which is to help individuals and society to have access to affordable medicines.[1] In their 2018 report, this organization highlights the case of Humira, an antiinflammatory drug marketed by AbbVie Pharmaceuticals (global sales in 2017: US$ 18.427 billion[2]; projection for 2020: US$ 21 billion[3]). This drug for the treatment of arthritis and inflammatory disease can be accessible as a biosimilar in many European countries; however, in the US that is not the case, as this chapter is written. The reason, according to I-MAK, is that AbbVie has pursued several patenting strategies that delay competition and allow the company to keep prices high while extending its US monopoly on the drug for many more years (until 2023). Specifically, according to I-MAK, AbbVie has filed 247 patent applications for Humira in the United States, compared with 76 in Europe. Almost 50% of Humira's US patent applications were filed in the past 4 years, even though the drug has been on the market since 2002. From 2012 to 2016, as AbbVie layered on more patents, the company raised the price by 18% every year. Total spending on Humira by Medicare and Medicaid during this time increased 266%.[4]

The Core Model
ISBN 978-0-12-814293-6
https://doi.org/10.1016/B978-0-12-814293-6.00006-8

The case of AbbVie/Humira is not an isolated one: it is actually the bread and butter of the biopharmaceutical industry as a whole. In the aforementioned report on the industry, I-MAK estimated that the 12 top-selling drugs in the United States have, on average, 125 patent applications each, which allow them to have an average of 38 years of attempted patent protections. The prices of 11 of these best-sellers have risen an average of 80% since 2012. Other specific examples include one of the most-often prescribed cancer treatments, Revlimid, a monoclonal antibody approved by the FDA in 2005 for the treatment of several hematological malignancies, which has 96 patents providing potentially 40 years without competition. Another example is Lantus, a treatment for diabetes, which may not see a generic alternative for 37 years due to the 49 patents issued. Moreover, the list continues.[5]

As I have described in *The World's Health Care Crisis* and in this book, the United States spends, in absolute terms, more than any other country in the world on pharmaceuticals due in large part to the fact that pharmaceutical companies charge to the consumers as much as they want. Unfortunately, this situation has severe consequences in other countries as well—accelerating the collapse of the global health care systems.

How can the Core Model help the situation, if at all? To answer this question let us examine the current IP system in the United States.

Patents

The origins of patents can be traced back, at least, to Ancient Greece, in 500 BC, in the Greek state of Sybaris (now Southern Italy), where it became possible for creators of unique culinary dishes to obtain exclusive rights for their recipes for 1 year. Fast-forwarding 1000 years, this type of legal protection assumed greater importance in 15th-century Venice, where inventive devices, especially those related to the autochthonous glass-blowing industry, had to be communicated to the Republic to obtain legal protection (for a period of 10 years) against potential infringers (*We can already see here the use of the Core Model, in which the Core* [*glass-blowing enterprise*] via *the Republic's legal system* [*Bridge*] *was protected to benefit the industry and, therefore, the Republic* [*Periphery*]). Over time, this system was implemented and consolidated in other parts of Italy, such as Florence, and in other European states such as England and France.

Moving forward in time, in the United States, the first patent was granted in 1790 after the first Patent Act of the U.S. Congress titled

"An Act to promote the progress of useful Arts",[6] was passed on April 10, 1790, which defined the subject matter of a US patent as "any useful art, manufacture, engine, machine, or device, or any improvement thereon not before known or used."[7] On July 31, 1790, Samuel Hopkins was the recipient of the first US-granted patent for a method of producing a substance called "potash" (potassium carbonate). This law was first revised and passed in 1793; it went through further major revision and passage in 1836, at which point this law instituted a significantly more rigorous application process, including the establishment of an examination system. The patent system certainly created a boom: it is estimated that between 1790 and 1836, about 10,000 patents were granted[8] and many more thousands were to come in the ensuing years.

According to the World Intellectual Property Organization, a patent is "an exclusive right granted for an invention, which is a product or a process that provides, in general, a new way of doing something, or offers a new technical solution to a problem. To get a patent, technical information about the invention must be disclosed to the public in a patent application."[9] In addition, there are three types of patents: the first one is a "utility patent," which may be granted to anyone who invents or discovers any new and useful process, machine, article of manufacture, or composition of matter, or any new and useful improvement thereof; the second type is a "design patent," which may be granted to anyone who invents a new, original, and ornamental design for an article of manufacture; and the third type is a "plant patent," which may be granted to anyone who invents or discovers and asexually reproduces any distinct and new variety of plant.[10]

In the pharmaceutical industry world, a "utility patent," which is the most common type of type of patent in this sector, can apply to one of four possible characteristics of a drug: the drug substance itself, the method of use, its formulation, or the process of making it. Drug-substance patents simply cover the chemical composition of the active ingredient; method-of-use patents cover the use of a drug in treating a particular condition such as heart failure or depression; formulation patents cover the physical form of a drug such as by mouth or injection; and process patents cover manufacturing methods. To be patentable, the "invention" is supposed to be "useful, novel, and nonobvious." *Useful* originally meant what it seems to—that it had some practical benefit. *Novel* meant that it was significantly different from earlier inventions, and *nonobvious* meant that it was not simply the next step that any knowledgeable person in the field would take, but rather a remarkably new concept.[11]

Today, in the United States, pharmaceutical companies are awarded 20 years of market exclusivity from the earliest date of filing with the U.S. Patent and Trademark Office (USPTO), for utility patents, once the patent is awarded. Due to the significant backlog of pending applications at the USPTO, the majority of newly issued patents receive some adjustment that extends the term for a period longer than 20 years.[12] To this number, one should discount the time it takes to develop a specific drug and to reach the market, which is approximately 12 years, on average.

Once a drug goes off-patent, the FDA allows the entrance to the market of generic versions of the drug—that is, a copy of the off-patent drug's active ingredient by other manufacturers. The Hatch–Waxman Amendments of 1984 enabled generics companies to launch a generic product simply by filing an Abbreviated New Drug Application (ANDA) that demonstrates that the generic product is bioequivalent to the brand drug that was already approved, and therefore companies did not have to replicate clinical trials. (The goal of this amendment of the Hatch–Waxman Act was to simplify the FDA generic-drug approval process for generics companies, the share of the market for which rose from less than 20% of prescriptions in 1984 to almost 85% of prescriptions in the present day, according to some US economists, though their sales are lower because these drugs are much cheaper. Besides generic drugs proper, there exists another line of generics called *brand generics*, which are medicines that have active ingredients that are *similar*, but not *identical,* to the active ingredients of the brand-name drugs that they imitate (therefore not infringing on patents), and which, for this reason, have the advantage of not having to go through clinical trials. These drugs are priced somewhere between brand-name drugs and true generics, and their market share is growing rapidly, but not without some controversy. Today, branded generics account for approximately 5% of the pharmaceutical market in the United States. As a result, within 2 years, brand-name sales erode as much as 80% of the original price. Because of their large size and the fact that the drug industry spends a great deal of resources and funding (amounting to more than 20% of its global sales) on marketing and promotion, companies need a huge amount of cash in revenue to compensate for losses due to patent expirations, which has become a heavy burden. Sometimes companies may choose, for strategic and financial reasons, to change a drug's status from prescription to over-the-counter (OTC).[13]

For biologics, becoming generic is a bit tough at present, because it is difficult to reproduce what is considered living material, although the

European Medicines Agency (EMEA) has approved a number of such products over the years (see Table 6.1). In the United States, the process has been much slower than in Europe. However, the FDA has seen a steady increase in biosimilar development programs since the passage of the Biologics Price Competition and Innovation Act of 2009 (BPCIA), with a number of biosimilars approved by the end of 2018 (see Table 6.2).

Long before a company begins clinical trials, companies protect the fruits of their Research and Development (R&D) through patenting and secrecy, because once clinical trials begin, it is not possible to withhold that information from the public. A factor that in the last decade has exacerbated the aggressive attitude by the pharmaceutical industry on patents is the patent cliff that the industry experienced during this period of time as well as the deficient pipelines. Let us delve into this for a moment.

Economic interests and patent cliff

Although the major challenge of the biopharmaceutical industry is to bring novel drugs to the market, over the last decade it has struggled—in the sense of losing some important sources of revenues—with the massive number of patent expirations of the industry's best-selling drugs. The impact of expiring patents are estimated to have cost between US$ 120 billion between 2009 and 2014,[14] and US$ 215 billion between 2015 and 2020, of which US$ 31 billion are at risk in 2018 alone.[15] Earlier, we discussed how the sales of off-patent products erode quickly within the first 2 years after a given patent has expired. Because of their large size and the fact that the drug industry spends a great deal of resources and funding (amounting to more than 30%–35% of its global sales) on marketing and promotion, companies need a huge amount of cash in revenue to compensate for losses due to patent expirations, which has become a heavy burden. In past years, some of the world's biggest pharmaceutical firms such as Pfizer and GlaxoSmithKline have lost billions of dollars in revenues from the patent expiration on some of their bestselling blockbuster drugs. Many of these drugs were discovered in the early 1990s. For example, blood-thinner Plavix (Bristol–Myers Squibb), leukotriene receptor antagonist Singulair (Merck), angiotensin II receptor antagonist Diovan (Novartis), and cholesterol-lowering drug Lipitor (Pfizer), faced patent expiration dates falling between 2011 and 2015. Biologic blockbuster drugs discovered in the late 1990s, such as Rituxan (Genentech), Humira (Abbvie), Novolog (Novo Nordisk), Avastin (Amgen), Remicade (Janssen Biotech), and

Table 6.1 European Medicines Agency (EMA)-approved biosimilars.[a]

Product name	Active substance	Therapeutic area	Authorization date	Manufacturer/ company name
Abasaglar (previously Abasria)	Insulin Glargine	Diabetes	September 9, 2014	Eli Lilly/Boehringer Ingelheim
Abseamed	Epoetin alfa	Anemia Cancer Chronic kidney failure	August 28, 2007	Medice Arzneimittel Pütter
Accofil	Filgrastim	Neutropenia	September 18, 2014	Accord Healthcare
Amgevita	Adalimumab	Ankylosing spondylitis Crohn's disease Juvenile rheumatoid arthritis Psoriasis Psoriatic arthritis Rheumatoid arthritis Ulcerative colitis	March 22, 2017	Amgen
Benepali	Etanercept	Axial spondyloarthritis Psoriatic arthritis Plaque psoriasis Rheumatoid arthritis	January 14, 2016	Samsung Bioepis
Bemfola	Follitropin alfa	Anovulation (IVF)	March 24, 2014	Finox Biotech
Binocrit	Epoetin alfa	Anemia Chronic kidney failure	August 28, 2007	Sandoz
Biograstim	Filgrastim	Cancer Hematopoietic stem cell transplantation Neutropenia	September 15, 2008 Withdrawn on December 22, 2016	CT Arzneimittel

Blitzima	Rituximab	Non-Hodgkin lymphoma Chronic B-cell lymphocytic leukemia	July 13, 2017	Celltrion
Cyltezo	Adalimumab	Crohn's disease Hidradenitis suppurativa Juvenile idiopathic arthritis Psoriasis Psoriatic arthritis Rheumatoid arthritis Ulcerative colitis Uveitis	November 10, 2017	Boehringer Ingelheim
Epoetin alfa Hexal	Epoetin alfa	Anemia Cancer Chronic kidney failure	August 28, 2007	Hexal
Erelzi	Etanercept	Ankylosing spondylitis Juvenile rheumatoid arthritis Psoriasis Psoriatic arthritis Rheumatoid arthritis	June 27, 2017	Sandoz
Filgrastim Hexal	Filgrastim	Cancer Hematopoietic stem cell transplantation Neutropenia	Feburary 6, 2009	Hexal
Filgrastim ratiopharm	Filgrastim	Cancer Hematopoietic stem cell transplantation Neutropenia	September 15, 2008 Withdrawn on April 20, 2011	Ratiopharm

Continued

Table 6.1 European Medicines Agency (EMA)-approved biosimilars.[a]—cont'd

Product name	Active substance	Therapeutic area	Authorization date	Manufacturer/ company name
Flixabi	Infliximab	Ankylosing spondylitis Crohn's disease Psoriatic arthritis Psoriasis Rheumatoid arthritis Ulcerative colitis	May 26, 2016	Samsung Bioepis
Grastofil	Filgrastim	Neutropenia	October 18, 2013	Apotex
Halimatoz	Adalimumab	Ankylosing spondylitis Hidradenitis suppurativa Juvenile rheumatoid arthritis Psoriatic arthritis Psoriasis Rheumatoid arthritis Uveitis	CHMP positive opinion June 1, 2018	Sandoz
Hefiya	Adalimumab	Ankylosing spondylitis Hidradenitis suppurativa Juvenile rheumatoid arthritis Psoriasis Uveitis	CHMP positive opinion June 1, 2018	Sandoz
Hulio	Adalimumab	Ankylosing spondylitis Crohn's disease Hidradenitis suppurativa Psoriasis Psoriatic arthritis Rheumatoid arthritis Ulcerative colitis Uveitis	CHMP positive opinion July 26, 2018	Mylan/Fujifilm Kyowa Kirin biologics

Hyrimoz	Adalimumab	Ankylosing spondylitis Crohn's disease Hidradenitis suppurativa Juvenile rheumatoid arthritis Papulosquamous skin disease Psoriatic arthritis Rheumatoid arthritis Ulcerative colitis Uveitis	CHMP positive opinion June 1, 2018	Sandoz
Herzuma	Trastuzumab	Early breast cancer Metastatic breast cancer Metastatic gastric cancer	February 14, 2018	Celltrion Healthcare
Imraldi	Adalimumab	Ankylosing spondylitis Arthritis Crohn's disease Hidradenitis suppurativa Psoriatic arthritis, Psoriasis Rheumatoid arthritis Ulcerative colitis Uveitis	August 24, 2017	Samsung Bioepis
Inflectra	Infliximab	Ankylosing spondylitis Crohn's disease Psoriatic arthritis Psoriasis Rheumatoid arthritis Ulcerative colitis	September 10, 2013	Hospira
Inhixa	Enoxaparin sodium	Venous thromboembolism	September 15, 2016	Techdow Europe

Continued

Table 6.1 European Medicines Agency (EMA)-approved biosimilars.[a]—cont'd

Product name	Active substance	Therapeutic area	Authorization date	Manufacturer/ company name
Insulin lispro Sanofi	Insulin lispro	Diabetes mellitus	July 19, 2017	Sanofi—Aventis
Kanjinti	Trastuzumab	Early breast cancer Metastatic breast cancer Metastatic gastric cancer	CHMP positive opinion March 28, 2018	Amgen/Allergan
Lusduna	Insulin glargine	Diabetes	January 4, 2017	Merck (MSD)
Movymia	Teriparatide	Osteoporosis	January 11, 2017	Stada Arzneimittel
Mvasi	Bevacizumab	Breast neoplasms Fallopian tube neoplasms Non-small-cell lung carcinoma Ovarian neoplasms Peritoneal neoplasms Renal cell carcinoma	January 15, 2018	Amgen
Nivestim	Filgrastim	Cancer Hematopoietic stem cell transplantation Neutropenia	June 8, 2010	Hospira (Pfizer)
Omnitrope	Somatropin	Pituitary dwarfism Prader—Willi syndrome Turner syndrome	April 12, 2006	Sandoz
Ontruzant	Trastuzumab	Early breast cancer Metastatic breast cancer Metastatic gastric cancer	November 15, 2017	Samsung Bioepis
Ovaleap	Follitropin alfa	Anovulation (IVF)	September 27, 2013	Teva pharma
Pelgraz	Pegfilgrastim	Neutropenia	CHMP positive opinion July 26, 2018	Accord Healthcare

Ratiograstim	Filgrastim	Cancer Hematopoietic stem cell transplantation Neutropenia	September 15, 2008	Ratiopharm
Remsima	Infliximab	Ankylosing spondylitis Crohn's disease Psoriatic arthritis Psoriasis Rheumatoid arthritis Ulcerative colitis	September 10, 2013	Celltrion
Retacrit	Epoetin zeta	Anemia Autologous blood transfusion Cancer Chronic kidney failure	December 18, 2007	Hospira
Ritemvia	Rituximab	Wegener granulomatosis Microscopic polyangiitis Non-Hodgkin Lymphoma	July 13, 2017	Celltrion
Rituzena (previously Tuxella)	Rituximab	Wegener granulomatosis Microscopic polyangiitis Non-Hodgkin Lymphoma Chronic B-cell lymphocytic leukemia	July 13, 2017	Celltrion
Rixathon	Rituximab	Chronic B-cell lymphocytic leukemia Microscopic polyangiitis Non-Hodgkin Lymphoma Rheumatoid arthritis Wegener granulomatosis	June 19, 2017	Sandoz

Continued

Table 6.1 European Medicines Agency (EMA)-approved biosimilars.[a]—cont'd

Product name	Active substance	Therapeutic area	Authorization date	Manufacturer/ company name
Riximyo	Rituximab	Chronic B-cell lymphocytic leukemia Microscopic polyangiitis Non-Hodgkin Lymphoma Rheumatoid arthritis Wegener granulomatosis	June 15, 2017	Sandoz
Semglee	Insulin glargine	Diabetes	March 28, 2018	Mylan
Silapo	Epoetin zeta	Anemia Autologous blood transfusion Cancer Chronic kidney failure	December 18, 2007	Stada Arzneimittel
Solymbic	Adalimumab	Ankylosing spondylitis Crohn's disease Hidradenitis suppurativa Psoriasis Psoriatic arthritis Rheumatoid arthritis Ulcerative colitis	March 22, 2017	Amgen
Terrosa	Teriparatide	Osteoporosis	January 4, 2017	Gedeon Richter
Tevagrastim	Filgrastim	Cancer Hematopoietic stem cell transplantation Neutropenia	September 15, 2008	Teva generics
Thorinane	Enoxaparin sodium	Venous thromboembolism	September 15, 2016	Pharmathen

Trazimera	Trastuzumab	Stomach Neoplasms Breast Neoplasms	CHMP positive opinion June 1, 2018	Pfizer
Truxima	Rituximab	Chronic lymphocytic leukemia Granulomatosis with polyangiitis Microscopic polyangiitis Non-Hodgkin's lymphoma Rheumatoid arthritis	February 17, 2017	Celltrion
Udenyca	Pegfilgrastim	Neutropenia	CHMP positive opinion July 26, 2018	ERA Consulting
Valtropin	Somatropin	Pituitary dwarfism Turner syndrome	April 24, 2006 Withdrawn on May 10, 2012	BioPartners
Zessly	Infliximab	Ankylosing spondylitis Crohn's disease Psoriatic arthritis psoriasis Rheumatoid arthritis Ulcerative colitis	May 24, 2018	Sandoz
Zarzio	Filgrastim	Cancer Hematopoietic stem cell transplantation Neutropenia	February 6, 2009	Sandoz

CHMP, Committee for Medicinal Products for Human Use; *VF*, in vitro fertilization.
[a]Data collected on May 12, 2011, updated on August 24, 2018.
Source: European Medicines Agency (EMA); and Generics and Biosimilars Initiative (GaBI); http://www.gabionline.net/Biosimilars/General/Biosimilars-approved-in-Europe.

Table 6.2 FDA-approved biosimilars and follow-on biologicals.[a]

Product name	Active substance	Therapeutic area	Authorization date	Manufacturer/ Company name
Admelog[b]	Insulin lispro	Diabetes	December 11, 2017	Sanofi
Amjevita (adalimumab-atto)	Adalimumab	Ankylosing spondylitis Crohn's disease Juvenile arthritis Psoriatic arthritis Psoriasis Rheumatoid arthritis Ulcerative colitis	September 23, 2016	Amgen
Basaglar[b]	Insulin glargine	Diabetes	December 16, 2015	Eli Lilly/ Boehringer Ingelheim
Cyltezo (adalimumab-adbm)	Adalimumab	Ankylosing spondylitis Crohn's disease Juvenile arthritis Psoriatic arthritis Psoriasis Rheumatoid arthritis Ulcerative colitis	August 25, 2017	Boehringer Ingelheim
Erelzi (etanercept-szzs)	Etanercept	Axial spondyloarthritis Polyarticular juvenile idiopathic arthritis Psoriatic arthritis Plaque psoriasis Rheumatoid arthritis	August 30, 2016	Sandoz

Fulphila (pegfilgrastim-jmdb)	Pegfilgrastim	Febrile neutropenia	June 4, 2018	Biocon/Mylan
Hyrimoz (adalimumab-adaz)	Adalimumab	Rheumatoid arthritis, juvenile idiopathic arthritis, psoriatic arthritis, ankylosing spondylitis, adult Crohn disease, ulcerative colitis, and plaque psoriasis	October 31, 2018	Sandoz
Inflectra (infliximab-dyyb)	Infliximab	Ankylosing spondylitis Crohn's disease Psoriatic arthritis Psoriasis Rheumatoid arthritis Ulcerative colitis	April 5, 2016	Pfizer (Hospira)
Ixifi (infliximab-qbtx)	Infliximab	Ankylosing spondylitis Crohn's disease Psoriatic arthritis Psoriasis Rheumatoid arthritis Ulcerative colitis	December 13, 2017	Pfizer
Lusduna[b] (tentative approval)	Insulin glargine	Diabetes	July 20, 2017	Merck
Mvasi (bevacizumab-awwb)	Bevacizumab	NSCLC Colorectal neoplasms Renal cell carcinoma Ovarian neoplasms Breast neoplasms	September 14, 2017	Amgen/ Allergan

Continued

Table 6.2 FDA-approved biosimilars and follow-on biologicals.[a]—cont'd

Product name	Active substance	Therapeutic area	Authorization date	Manufacturer/Company name
Nivestym (filgrastim-aafi)	Filgrastim	Autologous peripheral blood progenitor cell collection and therapy Bone marrow transplantation Cancer Myeloid leukemia Neutropenia	July 20, 2018	Pfizer (Hospira)
Ogivri (trastuzumab-dkst)	Trastuzumab	HER2 breast cancer HER2 metastatic gastric or gastroesophageal junction adenocarcinoma	December 1, 2017	Biocon/Mylan
Retacrit (epoetin alfa-epbx)	Epoetin alfa	Anemia (chronic kidney disease, Zidovudine, chemotherapy) Reduction of allogeneic red blood cell transfusions	May 15, 2018	Pfizer (Hospira)
Renflexis (infliximab-abda)	Infliximab	Ankylosing spondylitis Crohn's disease Psoriatic arthritis Psoriasis Rheumatoid arthritis Ulcerative colitis	April 21, 2017	Samsung Bioepis
Truxima (rituximab-abbs)	Rituximab	Relapsed or refractory, low-grade or follicular, CD20-positive, B-cell non–Hodgkin lymphoma (NHL) as a single agent	November 28, 2018	Celltrion/Teva

Udenyca (pegfilgrastim-cbqv)	Pegfilgrastim	Leukocyte growth factor indicated to decrease the incidence of infection, as manifested by febrile neutropenia, in patients with nonmyeloid malignancies receiving myelosuppressive anticancer drugs associated with a clinically significant incidence of febrile neutropenia	November 2, 2018	Coherus BioSciences, Inc.
Zarxio (filgrastim-sndz)	Filgrastim	Autologous peripheral blood progenitor cell collection and therapy Bone marrow transplantation Cancer Myeloid leukemia Neutropenia	March 6, 2015	Sandoz

NSCLC, Non-Small-Cell Lung Carcinoma.
[a]Data updated August 31, 2018.
[b]Admelog, Basaglar and Lusduna were approved via the FDA's abbreviated 505(b) (2) pathway as follow-on products not as biosimilars. No insulin lispro or glargine products were licensed under the Public Health Service Act at the time of filing, so there was no 'reference product' for a proposed biosimilar product.
Sources: Food and Drug Administration (FDA), https://www.fda.gov/drugs/developmentapprovalprocess/howdrugsaredevelopedandapproved/approvalapplications/therapeuticbiologicapplications/biosimilars/ucm580432.htm; GaBI Online, http://www.gabionline.net/Biosimilars/General/Biosimilars-approved-in-the-US; and companies' websites.

Herceptin (Genentech) have patent expiration dates between 2014 and 2020. Thus, patent expirations threaten to erode the revenue obtained through the sales of these products for their manufacturers and open up opportunities for other companies to generate revenue from selling biosimilar products. Nevertheless, we should not be surprised that their patents get somehow extended because biopharmaceutical companies spend significant resources in seeking routes to patent extensions, as mentioned at the beginning of this chapter.

To extend their patents companies develop new delivery methods for a given therapeutic, another strategy is to take a given generic company to court, delaying in that way the actual date of patent expirations. In addition, companies use a range of mitigation strategies that exist for life-cycle management including authorized generics, reformulation, pricing incentives, prescription to over-the-counter (OTC) switch, and pediatric exclusivity, etc.

It is estimated that in 2017, 18 branded drugs lost their patent protection, which represented annual sales for approximately US$ 26.5 billion in 2025. Among the 2017 expirations are: Roche's Rituxan, GSK's Advair, Eli Lilly's Humalog and Cialis, AstraZeneca's Byetta, Pfizer's Viagra, and Merck's Vytorin.[16] And this trend is expected to continue until at least 2025. For 2018, it is estimated that the losses are in the order of US$ 31 billion. Between 2015 and 2020, the losses to patent expirations could represent approximately $215 billion.[17] See Table 6.3 for potential patent expirations for 2018.

All of this creates a lot of growth pressure in the industry. However, we should also keep in mind that the patent protection of these products should end and that they should become generic. So, rather than focusing of "gaming the system" pursuing and applying "patent evergreening," pharmaceutical companies should focus on creating better strategies to come up with authentically novel drugs.

However, the issue of intellectual property is not only a problem of the industry; it affects academia as well, through the Bayh–Dole Act of 1980.

The Bayh–Dole Act

In the United States, the Bayh–Dole Act of 1980 (or Patent and Trademark Law Amendments Act (Public Law 96–517, December 12, 1980)[18] entitles US universities to the Intellectual Property Rights of discoveries made using federal funding. The rationale behind this act was to increase

Table 6.3 Expected patent expirations for 2018.

Drug brand name	Manufacturer	Indication	Global sales (billion US $), 2017	Notes
1. Lyrica	Pfizer	Medication for nerve and muscle pain	3.46	Patent protection is scheduled to expire in December 2018. Pfizer is reportedly working on getting an extension for pediatric exclusivity.
2. Rituxan	Genentech	Treatment for blood cancers and rheumatoid arthritis	7.9	Competition from biosimilars in Europe has already begun chipping away at sales for the drug, and its patent in the United States ended in 2018.
3. Cialis	Eli Lilly	Second only to Viagra in the market for erectile dysfunction medications	2.3	A settlement with generic drugmakers in 2107 extended its patent life; but expected to expire in September 2019. Teva launched a generic version of Cialis in September 2018.
4. Xolair	Novartis/Roche	Treatment for allergic asthma and chronic idiopathic urticarial	2.6	Patent is set to expire in 2018. No clear front-runners in the race to develop a biosimilar to challenge its place on the market.
5. Restasis	Allergan	Dry-eye treatment	1.41	Allergan made decided to license its patents for Restasis, a popular dry eye treatment, to the Saint Regis Mohawk Tribe, with the objective of that the company would be protected under tribal sovereignty laws. The move did not pay off. After a string of courts ruled against Allergan, Restasis, which sold $1.41 billion in sales in 2017, will begin facing competition from generics in 2018.

Continued

Table 6.3 Expected patent expirations for 2018.—cont'd

Drug brand name	Manufacturer	Indication	Global sales (billion US $), 2017	Notes
6. Advair	GlaxoSmithKline	Top-selling treatment for asthma and chronic pulmonary disease	4.17	Officially lost patent protection in 2010, no company has been able to get a generic version approved by regulators … yet.
7. Neulasta	Amgen	White cell booster	3.93	Patent ended in 2015, a biosimilar developed by Mylan is expected to enter the market soon.
8. Zytiga	Johnson & Johnson	Prostate cancer treatment	2.5	Generic threats from other companies are imminent.
9. Sensipar	Amgen	Treatment for secondary hyperparathyroidism in patients on dialysis	1.58	The company is still in several generic patent disputes, which could delay generics entry into market.
10. Ampyra	Acorda Therapuetics	Drug that helps improve muscle strength in patients with multiple sclerosis	0.54	This biotech company is now hoping that other medications in its pipeline, including a Parkinson's drug called Inbrija, will help make up for lost Amprya.

Adapted from Parrish, M., May 16, 2018. 10 Major Drugs Losing Patent Protections in 2018, Pharma Manufacturing https://www.pharmamanufacturing.com/articles/2018/10-major-drugs-losing-patent-protections-in-2018/; and other sources.

patenting of discoveries and the acceleration of economic growth through the creation of "high-tech" firms that license these technologies from the university.

With the Bayh—Dole Act, patenting, licensing, and material transfer agreements (MTAs) became an important part of translating basic innovation from university settings into commercial products. According to some legal scholars, this could have created a situation in which intellectual property constraints may have a negative effect on scientific communication and the use of such discoveries, described as knowledge anticommons or "the Tragedy of the Anticommons." This is a type of coordination breakdown, first expounded by Michael Heller in 1998,[19] in which a single resource has numerous rights-holders who prevent others from using it, frustrating what would be a socially desirable outcome. In a 1998 *Science* magazine article,[20] Heller and Rebecca Eisenberg, while not disputing the role of patents in general in motivating invention and disclosure, argue that biomedical research was one of several key areas in which competing patent rights could actually prevent useful and affordable products from reaching the marketplace. In fact, Intellectual Property Rights, although doubtless necessary, can be an impediment to the process of translating basic science into commercial products. However, of course, most companies and investors will not finance translational science and develop drugs without a strong intellectual property position to protect their investment.

The Core Model as exemplified by the development of bortezomib demonstrates that the "Tragedy of the Anticommons" can be solved via effective collaboration, not only within Heller's "Zone of Cooperative and Market-Based Solutions" (see Fig. 6.1),[21] but within the entire property

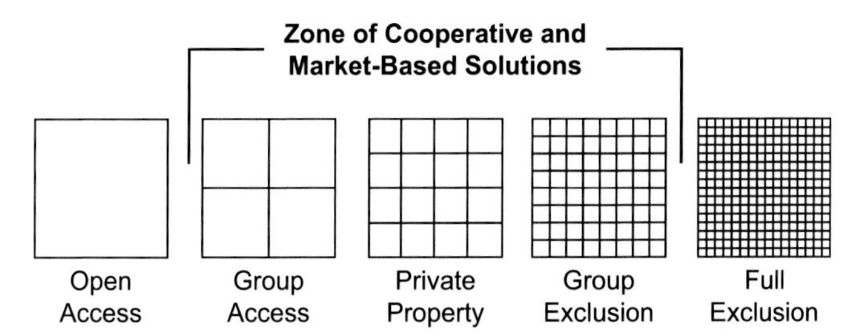

Figure 6.1 The full spectrum of property, revealed (Heller, 1999). *(Source: Heller, M.A., 1999. The Tragedy of the Anticommons. The Wealth of the Commons. Essay adapted from Chapter 2 of The Gridlock Economy (2010); http://www.gridlockeconomy.com.)*

spectrum, because even "closed" systems are interdependent and require, according to the Core Model, cooperation (Core, Bridge, and Periphery).

Instead of being IPRs an impediment for knowledge, the Core Model proposes that this situation could act actually as an incentive to develop new forms and methods to generate knowledge that could greatly contribute to the discovery and development of novel drugs. However, some ethical and legal rules have to follow. Let us take for example material transfer agreements, which many consider the bane of everyone in the technology-transfer business. Even though they may take time, they are necessary. For instance, someone could come in with a disclosure of his invention, and he used an antibody to create the invention, which probably came from somewhere else. Therefore, it is necessary to have some written agreement on who gets to use what material for what purposes. This is actually a fair thing to do.

In some cases, materials are exchanged freely between academia and industry, but at other times, the donor considers the transfer as proprietary material. Though the donor may be aware that the material is important for further research, the donor may allow its use under the condition that the donor is informed of whatever inventions are made when the transferred material is used and that the donor is given the right of first refusal. Therefore, universities need to have a written agreement in which they get permission from the company to use the material, or vice versa.

The Core Model and Intellectual Property Rights (IPRs)

Coming back to the question of how the Core Model can contribute to a mitigation of the inherent negative effects that come because of IPRs, there are different forms in which the model can help. One of them has been fully exemplified by the exchange of assets in the story case of the development of bortezomib, via collaborative interaction between start-ups, academia, Big Pharma, and the Periphery. Other paths are discussed subsequently, particularly in the Periphery.

The Periphery and IPRs

Given that one of the important factors contributing to the high cost of medicines worldwide is the "patent evergreening" strategy by the pharmaceutical industry, some individuals within the Periphery have taken the task to challenge this situation. This is the case of the already mentioned not-for-profit organization called I-MAK. According to this institution pharmaceutical companies, in general, apply for unmerited patents to block competition.[22] When these and other anticompetitive strategies are successful, a company can corner the market, artificially inflate the price of treatment, and block access to affordable medicines for decades. The solution that this not-for-profit organization provides is to act as "patent detectives" worldwide and in that way challenge unmerited patents worldwide. As a result, according to their website, they have been able, over the past 10 years, to increase access in 49 countries to 20 therapies for 8 diseases, including hepatitis C, human immunodeficiency virus (HIV), leukemia, tuberculosis, diabetes, cancer, and blood-related disorders. Worldwide, for HIV treatments alone, I-MAK reports that it has helped government health programs save more than $1 billion in the last 12 years. Similarly, in 2018, I-MAK successfully removed unmerited patents on Solvadi, the backbone of hepatitis C treatment. This victory for the 10 million people living with hepatitis C in China could save $50 billion to the health system if only 59% of infections are treated (similar to treatment rate for HIV).

Because of I-MAK's successful patent challenges on four HIV drugs in India, prices for these lifesaving generic medicines are now 51%–89% lower than the branded versions. These wins supplied low- and middle-income countries worldwide with low-cost generics, saving an estimated $500 million—money that could be reinvested to treat more than one million people.

I-MAK has worked alongside five patient advocacy organizations: International Treatment Preparedness Coalition (ITPC; Botswana),[23] Fundación Grupo Efecto Positivo (Foundation Positive Effect Group [FGEP]; Argentina),[24] Associação Brasileira Interdisciplinar de AIDS (the Brazilian interdisciplinary AIDS association [ABIA]; Brazil),[25] AIDS Access Foundation (Thailand),[26] and The All-Ukrainian Network of People Living with HIV (PLHIV)[27] since 2014 to reduce drug prices in Argentina, Brazil, Thailand, and Ukraine. Patent challenges, drug pricing negotiations, and policy dialogs between the public and private sectors on 15 HIV drugs in these four countries have contributed to price reductions, which are saving health programs an estimated $472 million each year.

Therefore, by acting as "patent detectives," I-MAK has been instrumental in removing barriers to treatment for some of the world's deadliest diseases such as hepatitis C and HIV. I-MAK helps patients, consumers, governments, and patent offices to create systems that support a competitive market in which the needs of patients and payers come first. The Core activities of I-MAK mission are:

- Providing evidence-based, data-driven research and analysis
- Fighting to reform patent policy worldwide
- Educating the public in global communities
- Filing patent challenges to get millions of people lifesaving drugs
- Defining global strategies and new solutions

It is very interesting to see how members of the Periphery, such as I-MAK and the ones we will see later and in the next chapters, behave like Core to achieve their goals!

Another initiative is the one recently taken by US billionaire John D. Arnold. Arnold is investing part of his fortune to campaign against high drug prices in the United States. Since 2014, he has spent more than $100 million in grants related to health care and has pledged $5.7 million to and I-MAK, which files legal challenges to the validity of certain US drug patents, to clear the way for lower-cost generics.[28]

We can rest assured that the pharmaceutical industry is very unhappy with these initiatives, which, if adopted worldwide, could become their real "nemesis."

Other activities, also stemming from the Periphery, further demonstrate how the Core Model can mitigate the effects of IPRs. Examples comprise the creation and success of a number of consortia in different parts of the world working at the preclinical and precompetitive level exploring a wide range of areas such as toxicogenomics, preclinical and clinical biomarkers, gene and protein annotation, proteomics, metabolomics, pharmacogenetics, etc. Successful examples include:

- **Biomarkers Consortium**[29]: This is an initiative created by the National Institutes of Health (NIH), the U.S. Food and Drug Administration (FDA), and the Pharmaceutical Research and Manufacturers of America (PhRMA), under the auspices of the Foundation of the NIH, focused on the development of biomarkers used in clinical research and drug discovery and development.
- **The Predictive Safety and Testing Consortium (PSTC)**[30]: Launched in 2006 by the not-for-profit Critical Path Institute, PSTC brings together pharmaceutical companies to share and validate innovative safety testing methods under advisement of the FDA (U.S. Food and Drug Administration), its European counterpart, the EMA (European Medicines Agency), and PMDA (Japanese Pharmaceutical and Medical Devices Agency).

- **Single Nucleotide Polymorphism (SNP) Consortium**[31]: The SNP Consortium (TSC) was established in 1999 as a collaboration of several companies and institutions to produce a public resource of single-nucleotide polymorphisms (SNPs) in the human genome. The initial goal was to discover 300,000 SNPs in 2 years, but the final results exceeded this.
- **Serious Adverse Event (SAE) Consortium**[32]: The international Serious Adverse Events Consortium (iSAEC) is a not-for-profit organization that coordinates and funds efforts to identify genetic factors that confer a risk for serious, drug-induced adverse events. The hypothesis driving this work is that genetic variation in some people predisposes them to serious adverse reactions to marketed drugs. By identifying DNA markers shared by patients who have experienced adverse reactions, the ultimate goal of the consortium is to improve the biomedical community's ability to predict which patients are at higher risk when taking specific medications, and to tailor treatment options accordingly. This information could also be helpful in designing safer drugs. According to their website, the iSAEC is at present developing novel, international clinical networks to increase their understanding of the genetics of the following SAEs across a diverse range of ethnic populations in different parts of the world:
  - Hepatotoxicity (DILI)
  - Serious Skin Rash (DISI)
  - Acute Hypersensitivity Syndrome (DRESS)
  - Nephrotoxicity (DIRI)
  - TdP/PQT effects (DITdP)
  - Excessive Weight Gain (associated with Class 2 Antipsychotic Medications)
  - IBD Therapy-related SAEs (four different phenotypes), and
  - Jaw Osteonecrosis (ONJ)

Another strategy currently being used is Patent Pools, which are a form of Open Source Collaboration and Innovation or consortia in which two or more patent owners agree to license one or more of their patents to one another or to third parties in areas associated with complex technologies that require complementary patents to provide efficient technical solutions.[33] Patent pools are beginning to be used to stimulate a wide range of fields. In biomedical research, they are being implemented in the area of neglected diseases, allowing both access to technologies and competitive business practices.[34] To illustrate this point, let us take the example of the collaboration between two private not-for-profit organizations: the Australian research institute called Cambia[35] and the Indian Open Source Drug Discovery (OSDD)[36] consortium. Cambia's main asset in this collaboration is its initiative Biological Innovation for Open Society (BiOS),[37] with the objective of creating "protected commons" that allows participants to use, modify, improve, and share existing technologies without infringing on proprietary rights. This is done through BiOS licenses, which follow the model of Creative Commons licenses commonly used online for multiple cultural purposes such as documents, photography, blogs, or General Public Licenses (GPLs) used in software development, etc. OSDD, on the other hand, is a consortium launched by India's Council of Scientific and Industrial Research (SIR) in 2007 with the objective of achieving "affordable health care through a platform in which talented minds can collectively discover novel therapies, as well as bring openness and collaboration to the drug discovery process, and keep drugs cost low."[38] By incorporating a web-based portal that allows participants access to bioinformatics tools, biological information, data on pathogens, and discussion forums, ODDS aims at breaking down drug discovery into smaller activities with clear deliverables. Participants can contribute anything that helps solve these problems and achieve the goals of these smaller activities, including ideas, software, articles, and intellectual property.

An important condition of OSDD is that users must grant worldwide nonexclusive rights to OSDD for the use of any IP rights acquired. In this way, it guarantees the overall collaboration of all users, while still protecting the Intellectual Property Rights of all involved. The collaboration between OSDD and its partners has created a renewed interest in focusing on solutions for tropical diseases like tuberculosis, malaria, and leishmaniasis, which take a large toll on the developing world, especially in the poorer countries.

Another important approach to mitigate the IPRs issue is drug discovery and development as carried out by not-for-profit research institutions such as the Stanford Research Institute (SRI),[39] founded in 1946, based in Menlo Park, California, now with 20 locations all over the world and 2100 employees, and more than 4000 patents in its roster. By reinvesting all of its profits in internal research and development and by collaborating with other major institutions such as the NIH, NSF, other government agencies, universities, and other organizations, SRI has advanced new therapies and vaccines to human clinical trials. For SRI the best collaborations are "mutually flexible, seek win/win solutions, leave their egos at home, listen first to understand, and play on each other's strengths."[40] SRI collaborative attitude is wide-ranging: from fellowships and innovation workshops to in- or out-licensing; to philanthropy outsourcing of research and development to partnering on funding proposals.

Other strategies include platforms such as Collaborative Drug Discovery Vault,[41] through which researchers can organize chemical structures and biological study data, and collaborate with internal or external partners through an easy-to-use web interface.

All of these models of collaboration are examples of the Core Model and work through knowledge transfer, integration, and translation. They all can create enormous generation of knowledge as long as there is goodwill and trust, and can use the resources of the Periphery and the Bridge to great effect. Therefore, collaboration can certainly trump the limitations created by IPRs.

In the next chapter, we will explore how the industry can implement the Core Model to great effect.

Endnotes

1. See I-MAK webpage: http://www.i-mak.org/mission/.
2. Abbvie News Center, 2018. AbbVie Reports Full-Year and Fourth-Quarter 2017 Financial Results, January 26. https://news.abbvie.com/news/abbvie-reports-full-year-and-fourth-quarter-2017-financial-results.htm.
3. Matthias, T., Mitra, A.K., 2017. AbbVie says Humira sales will balloon to $21 billion in 2020, shares rise, October 27, 2017. https://www.reuters.com/article/us-abbvie-results/abbvie-says-humira-sales-will-balloon-to-21-billion-in-2020-shares-rise-idUSKBN1CW1JK.
4. Amin, T., 2018. Commentary: Abuse of U.S. patent system drives high drug prices, Modern Health Care, October 27. https://www.modernhealthcare.com/article/20181027/NEWS/181029947.
5. Komendant, E., 2018. Pharmaceutical Patent Abuse: To Infinity and Beyond! Association for Accessible Medicines. https://accessiblemeds.org/resources/blog/pharmaceutical-patent-abuse-infinity-and-beyond.
6. Online at Library of Congress: "A Century of Lawmaking for a New Nation: U.S. Congressional Documents and Debates, 1774–1875": First Congress, Session II, chapter VII, 1790: An Act to promote the progress of useful Arts. http://memory.loc.gov/cgi-bin/ampage?collId=llsl&fileName=001/llsl001.db&recNum=232.
7. The U.S. Patent System Celebrates 212 Years. The U.S. Patent and Trademark Office. April 9, 2002. https://www.uspto.gov/about-us/news-updates/us-patent-system-celebrates-212-years.
8. Chartrand, S., 2004. Patents; The earliest U.S. patents went up in smoke. But a few are still being recovered, even 168 years after the fire. The New York Times, August 9. https://www.nytimes.com/2004/08/09/business/patents-earliest-us-patents-went-up-smoke-but-few-are-still-being-recovered-even.html.
9. World Intellectual Property Organization. https://www.wipo.int/patents/en/.

10. United States Trade and Patents Office (USTP). General information concerning patents. https://www.uspto.gov/patents-getting-started/general-information-concerning-patents.
11. See endnote 10.
12. See https://www.maxval.com/Introduction-to-patent-term-adjustment.html.
13. Mahecha, L., 2006. Rx-to-OTC switches: trends and factors underlying success. Nature Reviews Drug Discovery 5, 380–386.
14. VanEck, 2016. Drug patent expirations: $190 billion is up for grabs. Market Realist March 3, 2016. https://marketrealist.com/2016/03/drug-patent-expirations-190-billion-sales-grabs.
15. Parrish, M., 2018. 10 Major Drugs Losing Patent Protections in 2018. Pharma Manufacturing 16 May. https://www.pharmamanufacturing.com/articles/2018/10-major-drugs-losing-patent-protections-in-2018/.
16. Sagonowsky, E., 2017. Big Pharma faces $26.5B in losses this year as next big patent cliff looms, analyst says. FierceFarma, April 21. https://www.fiercepharma.com/pharma/big-pharma-faces-26-5b-patent-loss-threats-year-analyst-says.
17. See endnote 15.
18. See Public Law 96–517, December 12, 1980. https://history.nih.gov/research/downloads/pl96-517.pdf.
19. Heller, M.A., 1998. The Tragedy of the Anticommons: property in the transition from marx to markets. Harvard Law Review 111 (3), 621–88.
20. Heller, M.A., Eisenberg, R.S., 1998. Can patents deter innovation? The anticommons in biomedical research. Science May 01, 280 (5364), 698–701. http://science.sciencemag.org/content/280/5364/698.
21. Heller, M.A., 1999. The Tragedy of the Anticommons. The Wealth of the Commons; http://wealthofthecommons.org/essay/tragedy-anticommons. *Essay adapted from Chapter 2* of The Gridlock Economy *(2010);* http://www.gridlockeconomy.com.
22. See endnote 1.
23. See ITPC's website: http://itpcglobal.org/.
24. See the FGEP's website: http://fgep.org/es/.
25. See ABIA's website: http://abiaids.org.br/sobre-nos.
26. See AAF's website: https://www.changemakers.com/ashoka-fellows/entries/aids-access-foundation.
27. See All-Ukrainian Network of PLWH website: http://network.org.ua/en/.
28. Loftus, P., 2018. A billionaire pledges to fight high drug prices, and the industry is rattled, The Wall Street Journal October 21. https://www.wsj.com/articles/a-billionaire-decided-to-fight-high-drug-prices-and-the-industry-is-rattled-1540145686?tesla=y.
29. See the Biomarkers Consortium website: https://fnih.org/what-we-do/biomarkers-consortium.
30. See the PSTC website: https://c-path.org/programs/pstc/.
31. See: https://www.ncbi.nlm.nih.gov/pmc/articles/PMC165499/; https://www.genome.gov/10001884/single-nucleotide-polymorphisms-meeting/; https://www.researchgate.net/publication/233687285_Single_Nucleotide_Polymorphisms_SNPs_History_Biotechnological_Outlook_and_Practical_Applications.
32. See the iSAEC website: https://www.saeconsortium.org/.
33. WIPO, 2014. Patent Pools and Antitrust—A Comparative Analysis; https://www.wipo.int/export/sites/www/ip-competition/en/studies/patent_pools_report.pdf.
34. Chaguturu, R. (Ed.), 2014. Collaborative Innovation in Drug Discovery, Wiley, Hoboken, NJ.
35. See Cambia's website: https://cambia.org/.
36. See OSDD's website: http://www.osdd.net/.

37. See BiOS's website: http://www.bios.net/daisy/bios/home.html.
38. Massum, H. et al., Spring 2011. Open source biotechnology platforms for global health and development: two case studies. Information Technologies and International Development 7(1), 61–69. http://dev.itidjournal.org/index.php/itid/article/viewFile/697/295.
39. See SRI's website: https://www.sri.com/.
40. See endnote 34
41. See CDD Vault's website: https://www.collaborativedrug.com/.

CHAPTER 7

Global health care crisis, the pharmaceutical industry, and the Core Model

A health care crisis is primarily a *financial* crisis in which countries cannot successfully meet citizens' access to medicine due to the rising cost of health care services and, more importantly, pharmaceuticals. This crisis is worsened by age-dependency ratios and aging populations in these countries, and it then competes with the pension crisis for the money and political will that are needed to solve these problems.[1]

Not surprisingly, people across the planet constantly express dissatisfaction, frustration, anger, and despair about this situation, which is generally perceived as unsustainable in the near future and with no real or adequate solution envisioned on the horizon.

For a very long time, a health care crisis has been evident in the developing world and the poorer countries. However, over the last few decades, and particularly since the late 1980s and early 1990s, it has become evident in the United States (we just need to recall Hillary Clinton and President Bill Clinton's efforts in that direction) as well as in the rest of industrialized countries, including Europe. In Europe, which, as opposed to the United States, has a long tradition of institutionalized social welfare and where national health care systems were developed to create social safety nets for all their citizens, the situation has been very complicated, as I have described in *The World's Health Care Crisis*. In fact, health care costs in Europe have continued to rise in recent years, and, to keep costs low, a variety of payment and reimbursement systems (i.e., copayments, reference pricing, differential pricing, and others) have been created in addition to pharmaceutical price controls and shifted emphasis on the purchase of generic drugs. Other countries that have very socially conscious health care systems, such as Canada, Japan, and Australia, are struggling to find ways to deal with rising health and medicine costs.

The Core Model
ISBN 978-0-12-814293-6
https://doi.org/10.1016/B978-0-12-814293-6.00007-X

The US health care crisis

For many years, in the United States, many economists and public-policy makers have advocated for the implementation of a universal health care system in the United States based on the European and Canadian models, as we saw during the debates leading to the passing of the Affordable Care Act (ObamaCare) in 2010. This idea fails to recognize, however, that even universal health care coverage in itself is not a solution to the problem if facilitating access to affordable, safe, and effective medicines, which is the common denominator among all nations, is not met. As we live in consumerist and pharmaco-dependent societies, people expect and demand better medicines, and better, more-sophisticated medical interventions. This requires, as we have seen in the previous chapters, a great deal of innovation, labor, time, investment, and the development of very sophisticated technologies (including research platforms, biomarkers and diagnostics, imaging, etc.), which eventually also increase prices and overall costs. This is, unfortunately, a model similar to the one used by the telecommunications and automobile industries. If we want a more sophisticated smartphone, with higher memory and room for applications or "Apps," a better camera, and a more "chic" design; or if we want a car with a more sophisticated engine, more comfortable seats, and full extras, then it is obvious that we have to pay more. None of this is different from the biopharmaceutical industry's business at all!

It is, therefore, surprising that even today high-profile economists still believe that the US health care crisis is an issue of having health-insurance coverage when the concerning part, as I have demonstrated already,[2] is having access to the best drugs in an affordable manner. Not that the quality of health care services, attention, and preventive medicine programs are not important, quite the opposite! Nevertheless, they require political will. Rather, finding a cure and a treatment and, in some cases, even diagnosis for a given disease is unpredictable, difficult, expensive, risky, and complex. Therefore, we need to implement different and more-efficient approaches to the discovery and development of novel drugs. This is as true for the United States, which has the most privatized health care system in the world, as it is for Panama, Switzerland, England, Africa, China, Canada, Russia ... in fact, the entire world! If the United States were to adopt a universal health care coverage system (or its equivalent), wholly or mostly subsidized by the government, who would pay for more-expensive medicines to treat diseases such as cancer, for which treatment could cost more than US$100,000 a year for medicines alone? This does not even include the more-expensive drugs that are used to treat severe chronic ailments and rare disorders, which, as we have seen in Table 3.1, can cost over US$1 million a year (as in the case of Glybera [uniQure] for the treatment of lipoprotein lipase deficiency).This excludes other equally high costs, such as the costs of hospitalization, laboratory tests, and doctors' considerably high fees, which are in part due to the high cost of medical education and insurance, among other factors. Moreover, we have to remember that the purpose of not having drug price controls in the United States and having a privatized health-insurance system is, supposedly, to create incentives for the pharmaceutical and investment industries to take the high risks that are involved in producing more innovative drugs that people will pay via their private health insurance.

One actually wonders how the health care reform plan achieved by President Obama, the Affordable Care Act, or "ObamaCare," will survive, given its very shaky start and the ongoing attacks of the Trump Administration, including funding cuts for Medicare and Medicaid. But considering the fact that an ever-increasing number of patients are and will continue to be medicated with highly expensive drugs. It is often claimed that prescription drugs account for only 10%—15% percent of total health care costs. However, this is not a small number. In addition, it is only a statistical figure that does not take into account the individual socioeconomic and health circumstances of patients. In the United States, the percentage of people aged 50 or older has increased significantly, because the baby-boom generation has already reached that age, and statistical analyses have projected that from 2000 to 2030, the number of Americans over age 65 will double to around 71.5 million. Unfortunately, the very unhealthy lifestyle of the United States is increasing the incidence of many chronic disorders, such as diabetes, atherosclerosis, and hypertension, which will continue to increase health costs to the point of bleeding state and federal budgets and creating great disincentives for private companies to provide health-insurance coverage. The drugs used to treat such chronic conditions are very costly, and millions of people are struggling with these costs, not only in the United States but all over the world.

The pharmaceutical factor

Missing often in health care debates is the following question: How will people have access to the best medicines in the world, and who is going to pay for them?

For many health care economists and for many politicians in the United States, this is not a big problem given that at present 85% of prescriptions written in this country are for generic drugs, which are "easily" affordable—which, as we all know, is not true. If we think about this carefully, we realize that this has many limitations given that generics companies are usually not innovators, because they make money from drugs that have already been discovered and developed by other companies and that have gone off patent. Effective though these drugs may be, there is always room for improvement in terms of dosing, effectiveness, or safety. On the other hand, they may become cheaper in the United States, but not in the rest of the world after we consider purchasing-power parity and other important economic considerations.

However, there is another important reason: treatments are only available for less than 3%–5% of the numerous diseases that affect humankind (*Though these are contested figures, at present more than 12,000 classified diseases exist [of which 8000 are rare] and counting, and treatment exists for only 600 of them!*). These are extremely important unmet medical needs. Diseases such as those requiring antibiotics will continue to increase, not to mention rare/orphan, neglected, international infectious diseases. Another issue is the increasing incidence, diversity, and complexity of chronic ailments among the world population. Furthermore, regulators recently acknowledged that some generics really are not equivalent to the branded drugs that they are imitating, which has prompted the U.S. Food and Drug Administration (FDA) to consider (and adopt) tougher new standards to make sure "there is less variability" among generics, which has certainly concerned the generics industry. Not to mention the ever-increasing problem of fake and counterfeit drugs in the developing world and even in the United States!

So, what is the solution?

Health care reform, therefore, should be not only about health-insurance reform, but also about reform of *prevention*, biopharmaceutical industry, intellectual property law, regulatory system, university–industry collaboration, basic science and innovation, pharmaceutical marketing and pricing, tort, medical education, primary health attention and infrastructure, and other types—all of which contribute directly and indirectly to the cost of the health care system. Nevertheless, the most difficult challenge of all, as mentioned already, is drug discovery and development. To deal with these problems, we need a paradigm, such as the Core Model, that will bring multilateral collaboration, convergence of efforts, focus, cross-communication, and *delivery* on major issues related to health care.

Current state of the pharmaceutical industry

As I have said elsewhere, the pharmaceutical business model is, indeed, very simple: pharmaceutical companies create new products, launch them, and grow them over time, and, finally, these products go off patent. However, many pharmaceutical companies, as we have seen, are doing all they can to avoid the patent expiration of their best-selling drugs and block generic competition, even if this strategy is, at least morally and sometimes legally, very questionable.

The internal problems that the pharmaceutical industry has to confront today cover several areas, especially its scientific and commercial bases and the increasing approval hurdles posed by regulators. At the scientific level, the major problem is attrition. The high failure rate in drug development is very costly for the industry (only 1 in 10,000–50,000 potential candidates succeeds in late-stage drug approval, as mentioned in Chapter 3). Therefore, every new product that comes to the market has to cover the cost of failure of many other products. Unfortunately, companies are extremely secretive about their failures, so one never knows for sure what goes wrong in preclinical studies and clinical trials in other firms to avoid repeating the same mistakes or improve the way things are done in both the discovery and the development phases. Moreover, I wonder about the impact of this behavior across the industry and on society. In the end, the cost is passed on to the patient, the global health care systems, and the economy. Another problem is that since the year 2000, due to more knowledge about the biology of disease, better and more-refined technologies are used in drug discovery and development. This includes the way clinical trials are done (compared to 50 y ago), so that drugs often reach market in indications in which they are considered the best of their kind and, therefore, difficult to improve upon. It is difficult to improve on a statin such as Lipitor, so companies decided to develop drugs, such as Crestor, with combined modes of action. Thus, people may rightly say that the low-hanging fruits are gone, with the big indications now covered by very satisfactory drugs. Not that no improvements are possible for the treatment of hypertension or diabetes, but these areas are reasonably well covered for the general population—so much so that it has become difficult to come up with something that is significantly much better. This would require a major breakthrough, and breakthroughs occur infrequently in these areas. Therefore, the differences between effective drugs that are hitting previous targets and the next generation of drugs will greatly depend on our understanding of the genome and the relationship between diseased genes and phenotypes. That is going to take time. The pharmaceutical industry is at an inflection point, a threshold at which the genome has been mapped and at which scientists are beginning to understand the biology of the proteins that are encoded by the genome. We also now have powerful technologies, such as Clustered Regularly Interspaced Short Palindromic Repeat (CRISPR), which have great potential to transform medicine. Given this, one could expect a lag between the current efforts to understand the biology and the mechanisms of action and the approval of novel drugs based on this understanding. However, this research promises a large number of wonderful new drug introductions in the future, by which point both the biotechnical (biotech) and pharmaceutical industries will have changed dramatically. Nevertheless, I fear that there is not much to sustain growth.

From the commercial perspective, the industry is under constant growth pressure, due to the large size of its companies (see, for instance, Table 7.1 the size of the top 10 pharmaceutical companies), and constant need to achieve double-digit growth to maintain high market capitalization. In addition, because of the megamergers that occurred in the last couple of

Table 7.1 The top 10 pharmaceutical companies in the world based on global sales, 2017–2018.

Name of company	Global sales (US$ billion)
1. Pfizer	52.54
2. Roche	44.36
3. Sanofi	36.66
4. Johnson & Johnson	36.3
5. Merck	35.4
6. Novartis	33.0
7. Abbvie	28.22
8. Gilead	25.65
9. GlaxoSmithKline	24.0
10. Amgen	22.85

From Proclinical. https://www.proclinical.com/blogs/2018-3/the-top-10-pharmaceutical-companies-in-the-world-2018 *(original source).*

decades, pharmaceutical companies have become so large that it has become very difficult to manage them from the top.

From a regulatory perspective, companies are required to the demonstrate that what comes out of clinical trials is not only safe, but also better than what is already available in the market, and these are really high hurdles, especially for smaller biotech companies. Therefore, companies have to become very creative about how to design their clinical trials.

Despite the wave of mergers that is currently underway and the constant acquisition of many related technology platforms and potential drug candidates, the success of this industry will always lie in its Research and Development (R&D) productivity. Moreover, increased R&D productivity can be achieved by (1) lowering production costs and the time required for drugs to reach the market, (2) reducing the failure rates of leading compounds, and (3) reorganizing R&D infrastructure. In all of these aspects, the Core Model can be very effective, as we will see.

The Core Model

Prior to the Bayh–Dole Act, the pharmaceutical industry always kept a close relationship with academia. However, with Bayh–Dole, although pharmaceutical companies continued to have collaborative agreements with academia, the industry grew more "close" and "secretive" due to intellectual property issues and, perhaps, increased competition from the new biotechnology firms. Therefore, the pharmaceutical industry has had to adapt to this new source of competition. But a negative attitude began to be born in academia, in which the applied science being performed in industry was seen as the "dark" side, exclusively "money-oriented," scientifically "less-rigorous," and exploitative of academic basic (and usually open) research.

Since 2010 or so, with higher challenges to develop novel drugs, the threat of patent expirations, and the significant increment in investment, time, labor, regulatory requirements, and overall complexity of science, and new successful models implemented in other industries, such as "Open-Source Innovation"—which has characterized the software and electronic industries—the pharmaceutical industry has certainly become more open. In addition, sustained funding of the industry in academic basic research (with potential in applied and translational research) has contributed to a change in the academic mentality, which has resulted in a better understanding of the industry's goals as well as the materialization of mutually beneficial projects and technology-transfer agreements.

This increased opening in the pharmaceuticals (pharma) business model can be seen in its increased investment in collaboration with academia (as we shall see in the next chapter). This collaboration addresses such areas as early drug discovery, target identification and validation, probe development, stem-cell research, bioinformatics, big-data analytics, translational research, the developments of biologics and natural products, rare/orphan and tropical diseases, and clinical trials. In a bold move, as we will soon see, Big Pharma has opened its doors to public consortia and the public regarding access to complex data sets—with the goal of generating new and fresh insights, global expertise, and novel interpretation. All of this could be reflected in the creation of safer and more efficacious drugs, while saving time, labor, and capital. This approach is very much implementing the Core Model.

Some of the major recent research collaborations having these goals in mind include:

- **Merck and Harvard University:** These institutions have entered into an exclusive license and research collaboration agreement to further the development of small-molecule therapeutics for leukemia and other cancers. Under the terms of the license agreement, Merck will pay to Harvard an up-front fee of US$20 million and will be responsible for development, including clinical development, and for worldwide commercialization of products. The University is also eligible to receive development and commercialization milestone payments, as well as tiered royalties on any resulting products.[3]
- **AstraZeneca and Columbia University:** AstraZeneca's Center for Genomics Research (CGR) and Columbia's Institute for Genomic Medicine (IGM) have established a collaborative genomics alliance. This partnership will develop novel statistical genetics and computational methods, driving innovation in gene discovery from large whole exome and whole genome data sets. In addition, IGM has forged alliance with Biogen (gene discovery and novel therapeutic approaches to treat neurological disorders), Gilead (genetics, drug toxicity and efficacy in liver disease), University of California, Berkeley (UCB) (gene discovery and sequencing clinics), and Janssen Endoplasmic Reticulum (ER) Stress Disorders.[4] AstraZeneca, on its part, has recently established three partnerships, with Regeneron and the UK Biobank, the FinnGen consortium (a large-scale genome study of the Finnish

population), and the University of Cambridge, with the goal of uncovering the genetic drivers of some diseases.[5]

- **Pfizer:** Early in 2018, Pfizer announced the creation of a new partnering model for early stage academic research, called the Innovative Target Exploration Network (ITEN), with the goal of identifying academic research projects that have a potential to deliver innovative therapeutic targets and mechanisms of action within Pfizer's core areas of expertise. Among the universities that have partnered with Pfizer's key hubs of excellence are University of Cambridge, Oxford University, and University of Texas Southwestern. In 2010, Pfizer created the Centers for Therapeutic Innovation (CTI), a pioneering research and development network initiated by Pfizer that uses an open innovation model to bring great ideas to fruition by using translational science to create clinical candidates. Pfizer's facilities, in at least 20 places globally, including Massachusetts, New York, California, the UK, are conveniently located on or near academic campuses, allowing their partners to work closely with Pfizer scientists to translate research ideas into clinical applications, with the ultimate goal of moving a therapeutic hypothesis through Proof of Mechanism (PoM) in humans. Through CTI, Pfizer is departing from the historical pharma model of R&D and transforming the traditional model of drug discovery to solve key challenges—namely the high cost and substantial time investment—of drug discovery.[6] This is another clear example of the application of the Core Model.
- **Novartis:** Novartis and its research division, the Novartis Institutes for Biomedical Research (in Cambridge, MA) over the years has established important collaboration agreements with a large number of universities and institutions, including the Friedrich Miescher Institute in Basel, Switzerland, the Massachusetts Institute of Technology (MIT), University of California Berkeley, Harvard University, the Broad Institute, and many others. Novartis has more than 300 academic collaborations in its global research portfolio.[7] In addition, Novartis has recently established a number of partnerships with other pharmaceutical and biotech companies. For example, it developed an expanded commercialization agreement with Amgen for a compound investigated for the prevention of migraine[8]; and recently with Pfizer, one which will include a study combining Novartis drug tropifexor and one or more Pfizer compounds for the treatment of nonalcoholic steatohepatitis (NASH).[9]

- **Ely Lilly:** This company, which is one of the pioneers in pharmaceutical "Open Innovation," has created the Open Innovation Drug Discovery (OIDD) initiative, focused on collaborative efforts with external partners to push science forward. Within this model, Ely Lilly engages external investigators in a hypothesis-driven approach to early drug discovery. Lilly gives scientists access to the same tools and expertise available to their scientists to design, test, and make new molecules. However, the collaborator retains complete control of their intellectual property, while still benefiting from the proprietary analysis that Lilly provides. External collaborators can use Lilly's molecules to test their own biological hypotheses.[10,11] In 2009, Lilly created the Lilly Phenotypic Drug Discovery Initiative, or PD2 (pronounced PD-squared), which uses Lilly-developed disease-state assays and a secure web portal to evaluate the therapeutic potential of compounds synthesized in university and biotech laboratories. Findings from this initiative could ultimately form the basis for collaboration or licensing agreements between Lilly and external institutions in areas such as Alzheimer's disease, cancer, diabetes, and osteoporosis.[12]
- **Bayer:** Bayer has also jumped on the Open Innovation bandwagon with its initiative "Grants4Targets," which aims at fostering collaborations with academia and start-ups through a wide range of activities including codevelopment, funding, mentoring, partnering, licensing, etc.[13]
- **Apollo Therapeutics:** Created in 2016, this consortium comprising some world-leading UK Universities and three leading pharmaceutical companies (AstraZeneca UK Limited, Glaxo Group Limited, and Johnson & Johnson Development Corporation [JJDC], Inc.) created a £40 million fund to drive therapeutic innovation. Each of the three industry partner companies (AstraZeneca UK Limited, Glaxo Group Limited, and JJDC) will contribute £10 million over 6 y to the venture. The technology transfer offices (TTOs) of the three university partners—Imperial Innovations Group Plc, Cambridge Enterprise Ltd., and University College London (UCL) Business Plc—will each contribute a further £3.3 million. The aim of Apollo is to advance academic preclinical research from these universities to a stage at which it can either be taken forward by one of the industry partners following an internal bidding process or be out-licensed. The three industry partners will also provide R&D expertise and additional resources to assist with the commercial evaluation and development of projects.[14]

The list will likely continue. See Tables 7.2 and 7.3 for a summary of collaborations between academia and pharma in the last few years.

Furthermore, the pharmaceutical industry is expending ongoing effort to approach drug discovery through consortia in a way similar to the one that academic centers and not-for-profit organizations had in the past.

Table 7.2 Recent collaborations between pharma and academia to accelerate drug discovery.

Private sector	Public sector	Drug target research area
Agilent	UC, Berkeley	Synthetic biology institute
Astra Zeneca	Broad Institute	Infectious disease
Astra Zeneca	University of Manchester	Inflammatory disease
Astra Zeneca	University College London	Stem cells
Astra Zeneca	Weil Cornell Medical College, Washington School of Medicine, Feinstein Institute for Medical Research, University of British Columbia	Alzheimer's disease
Bayer	UCSF	R&D agreement (10 y)
Bristol–Myers Squibb	Vanderbilt University	Parkinson's disease
Gilead	Yale	Oncology (US$40 M)
GlaxoSmithKline	MD Anderson Cancer Center	Cancer immune therapy
GlaxoSmithKline	Harvard Stem Cell Institute	Heart disease and cancer (US$25 M)
GlaxoSmithKline	University College, London	Amyloidosis: Transthyretin protein stabilization
GlaxoSmithKline	Vanderbilt University	Obesity
GlaxoSmithKline	Emory Institute for Drug Discovery	Drug discovery for rare diseases such as malaria
GlaxoSmithKline	Yale University	Cancer, inflammation, infectious disease
Genentech	UCSF	Neurodegenerative diseases
Johnson&Johnson	MIT	Oncology early academic research
Johnson&Johnson	Sanford–Burnham	Alzheimer early academic research
Johnson&Johnson	Queensland University	Chronic pain

Continued

Table 7.2 Recent collaborations between pharma and academia to accelerate drug discovery.—cont'd

Private sector	Public sector	Drug target research area
Merck	California Institute for Biomedical Research	Preclinical proof of concept
Novartis	Harvard	Stem cells
Novartis	University of Pennsylvania	Center for Advanced Cellular Therapies (US$20 M)
Pfizer	University of Pennsylvania	Research, clinical development, and policy (US$I5 M)
Pfizer, Entelos	University of California, University of Massachusetts, MIT	Energy metabolism, diabetes, and obesity (US$14 M)
Pfizer	University College London	Stem cells based therapies
Pfizer	Washington University, St. Louis	Immunology, indications discovery
Sanofi Aventis	Columbia University	Diabetes
Sanofi Aventis	Salk institute	Gene therapy
Sanofi Aventis	Stanford	Stem cells
Sanofi Aventis	Harvard	Cancer, diabetes, inflammation
Sanofi Aventis	University of California	Diabetes
Takeda	Sanford–Burnham	Obesity
Veridex	Massachusetts General Hospital	Circulating tumor cells
Zambon	UCSF	Drug delivery

Source: Chaguturu, R., et al., 2014. Collaborative Innovation in Drug Discovery. Wiley.

Table 7.3 Recent drug discovery through pharma-academia consortia.

Private sector	Public sector	Drug target research area
AstraZeneca, Eli Lilly, GlaxoSmithKline, Janssen Pharmaceutica, Novartis, Orion, Pfizer, Roche, Servier, and Wyeth.	Karolinska Institute (Sweden), The University of Cambridge (United Kingdom), Central Institute of Mental Health (Germany), Consejo Superior de Investigaciones Científicas (CSIC) (Spain), University of Manchester (United Kingdom), and the Bar Ilan University (Israel).	Novel methods leading to New Medications in depression and Schizophrenia (NEWMEDS)

Table 7.3 Recent drug discovery through pharma-academia consortia.—cont'd

Private sector	Public sector	Drug target research area
Pfizer CTI	NYC, Boston, San Francisco University Consortium	Multiple areas
GlaxoSmithKline, Pfizer, Inc., and AstraZeneca	Critical Path Institute, a nonprofit partnership with the FDA	Coalition against major diseases to share data on thousands of Alzheimer's patients in hopes that the extra information will spark new ideas for treatments
Eli Lilly	Open-access submission of compounds	Phenotypic drug discovery platform
GlaxoSmithKline (GSK)	Open-access database	Genomic and protein expression profiling data for over 300 cancer cell lines via the NCI's Cancer Bioinformatics Grid for academia to mine
Merck-Sigma Advanced Genetic Engineering (*SAGE*) Labs	Open-access database	Sharing and disseminating complex data representing disease biology (genetic [single-nucleotide polymorphism (SNP), copy number variations], RNA expression [messenger (m)RNA, micro (mi)RNA, other noncoding RNA])
Pharma Consortium	CTSA Pharmaceutical assets Portal-NIH	To improve information exchange regarding drugs available for repurposing —proactively engage pharma in data sharing

Source: Chaguturu, R., et al., 2014. Collaborative Innovation in Drug Discovery. Wiley.

This is because the industry has now understood that the complexity of biological systems is just too great to be addressed by one single organization (see Table 7.4 for a short list of such consortia). These consortia have made the data public through different platforms and databases containing the results of pharma screening campaigns, ADME/tox (absortion,

Table 7.4 Consortia between Eli Lilly, GSK, Merck, and the Clinical and Translational Science Award portal.

1. *Merck-Sage Bionetworks.* This is an open-access, nonprofit, organization established by Merck,* which consists of data from human and mouse disease models from the Rosetta subsidiary of Merck. The public can access Merck data on systems-biology networks and can also download the essential computational disease-biology software tools for data access and analysis. The Sage program is an open-access approach designed to bring academia and pharma together to develop comprehensive human disease biology models to predict network biology in normal physiology and disease and for all sources to deposit integrated networks of biological data. This aims to facilitate the integration of diverse molecular megadata sets to build predictive bionetworks and to establish a precompetitive position for human disease biology. The long-term goal of network biology is aimed at designing better and more-targeted drugs based on systems-network models.
2. *GSK-caBIG Collaboration.* GSK released genomic profiling data for over 300 cancer cell lines via the National Cancer Institute's Cancer Bioinformatics Grid (caBIG), a network of infrastructure and tools that enables the collection, analysis, and sharing of data. The site provides valuable information for genomic profiles for a wide variety of tumors, including breast, prostate, lung, and ovarian cancers. The public access allows any researcher to download the open-source GSK cancer data through cancer microarry informatics (caArray).
3. *ChEMBL-Neglected Tropical Disease Archive.* Large volumes of primary screening and medicinal chemistry data targeting neglected diseases are available through ChEMBL—Neglected Tropical Disease Archive. †The open-source repository contains the structures and screening data against the malarial parasite from GSK (13,500 compounds), Novartis (>5600 compounds), and St. Jude's Children's Research Hospital (310,000 compounds). The data are available to all for annotation and analysis in the hope of expediting development of new therapeutics against neglected diseases.
4. *EU Innovative Medicines Initiative—Joint Undertaking* (IMI-JU). This initiative represents a public—private partnership between the European CCommunity and the pharmaceutical industry (the European Federation of Pharmaceutical Industries and Associations [EFPIAs]). The goal of the Safer And Faster Evidence-based Translation (SAFE-T) consortium‡ is to identify sensitive translational safety biomarkers for drug-induced kidney, liver, and vascular injury. The goal is to predict drug toxicity in humans by minimizing failures related to drug toxicity data from species differences in preclinical safety tests and nonrepresentative patient populations in clinical trials, The Integrating bioinformatics and chemoinformatics approaches for the development of expert systems allowing the in silico prediction of toxicities (eTOX) program of IMI aims to create a database of high-quality in vitro and in vivo data and predictive toxicological models. The data are available to pharma and academia and serves to highlight the ADME/tox

Table 7.4 Consortia between Eli Lilly, GSK, Merck, and the Clinical and Translational Science Award portal.—cont'd

issues associated with compounds and have the potential to greatly reduce the high attrition rates associated with many lead scaffolds.

5. *Innovative Medicine Initiative.* Europe's Innovative Medicine Initiative,** a public–private consortium, has committed US$271 million investment to support discovery of novel drug candidates. The consortium currently consists of 30 academic and corporate collaborators [23].
6. *Structural Genomics Consortium (SGC).* This is an open-access, not-for-profit, public–private partnership to contribute to basic biology relevant to drug discovery using large-scale 3-D structures of proteins of therapeutic importance from humans and their parasites. The SGC includes collaborations between the Universities of Toronto and Oxford and the Karolinska Institute in Stockholm with GSK, Eli Lilly, Pfizer, the Novartis Research Foundation, the Wellcome Trust, and Canadian agencies. The SGC has also collaborated with the chemistry and biochemistry departments of University of Oxford, the NIH Chemical Genomics Center in Washington, DC, and GSK to set up open-access chemical probes for epigenetic proteins.

* See http://sagebase.org
† See http://www.ebi.ac.uk/chembl/
‡ See http://www.imi-safe-t.eu
** See http://www.imi.europa.eu
Source: Chaguturu, R., et al., 2014. Collaborative Innovation in Drug Discovery. Wiley.

distribution, metabolism, excretion, and toxicology) studies, and genomic and expression arrays, etc., for interpretation, annotation, and further drug probe development or for diagnostics.[15] The rationale behind this strategy of data sharing and analysis between major pharmaceutical companies and academic researchers is two-fold: first, to attract and bring together the insight of global intellect, at no cost; second, to expedite the translation of complex data into a drug discovery pipeline. Though some people may criticize this strategy as a "free-ride," in reality, it can be of great benefit for academic scientists and institutions, small research facilities, not-for-profit organizations, small biotech companies, and so on. This is because most of these cores cannot, on their own, undertake the very costly activities of high-throughput biomarker development, molecular and metabolic studies in such a high quality and large scale. This approach allows these organizations to avoid cost- and time prohibitive research and instead redirect their efforts toward nonredundant areas of future therapeutic research. This is another exquisite example of the Core Model.

Other Core Model examples can be found in the field of drug discovery through repositioning and repurposing strategies in which academia and not-for-profit organizations are trying to achieve the same success that some biotech and pharmaceutical companies have had in finding new uses for old approved or abandoned drugs.[16,17] In this respect, the National Institutes of Health (NIH) is playing a fundamental role through the National Center for the Advancement of Translational Sciences (NCATS) program Clinical and Translational Science Awards (CTSA).[18] This program supports a national network of medical research institutions called "hubs" that collaborate with the objective of improving translational research, facilitating in this way the discovery of novel treatments. These hubs work together at the local and regional levels to catalyze innovation in areas such as training, research tools, and processes. One important strength of this program is to enable all the constituents of the Core Model (research teams of scientists, individual scientists, pharma, patient advocacy organizations, and community members) to work together in solving system-wide scientific and operational problems in clinical and translational research that no single constituent can overcome. This major contribution shortens the time it takes for innovative medicines to reach patients globally.

In terms of drug repurposing, one of the goals of the CTSA is to facilitate collaborations between pharma and the CTSA/NIH consortium, which, as mentioned previously, include academia and not-for-profit academic groups, to find new targets for discontinued or failed compounds that have been previously characterized. It is estimated that CTSA mediates one of the largest pharma–academia integration efforts involving 348 researchers nationwide, with diverse target-disease interests, and eight pharmaceutical companies (Pfizer, Merck, GlaxoSmithKline [GSK], Novartis, Genentech, Abbott, Eli Lilly, and AstraZeneca).[19] The University of California–Davis, which is one of the CTSA grant beneficiaries, has established created a Pharmaceutical Assets Portal. This portal allows academic researchers, entrepreneurs, not-for-profit organizations, and firms to gain access to compounds from pharma that were abandoned during clinical development, eliminating in this way barriers to access to these compounds[20] and to data that otherwise would be extremely difficult to obtain or gather independently.

Other approaches in the public domain exemplifying the Core Model are:

- **The Strategic Pharma-Academic Research Consortium (SPARC)[21]:** This is an NIH CTSA Program organization that fosters innovative collaborations between academic centers that form part of the NIH CTSA Program network and the pharmaceutical industry at the levels of discovery and "precompetitive" target discoveries. This is a platform for research that utilizes the unique strengths of academic and industry for basic discovery, target identification, and testing tool-molecule chemical biology and translational research. Started in 2014 by four CTSA Program hubs from the Midwest with Eli Lilly and Takeda Pharmaceuticals (both of which have a large presence in the Midwest), this consortium is housed in the Indiana Clinical and Translational Sciences Institute. SPARC's mission is "to establish a public-private partnership for patient-focused discoveries that generate greater knowledge and better approaches to next-generation and targeted therapies." Another goal is to utilize patient data and samples to obtain a better understanding of the physiology of a given disease. The following institutions form part of the consortium: in academia: Indiana University, Ohio State University, Northwestern University, University of Chicago, and Washington University in Saint Louis; in the pharmaceutical industry: Eli Lilly and Takeda. The consortium initially focused on autoimmune diseases as the major theme for selecting projects, but, over time, it has funded five projects and has incorporated other institutions, more than a dozen faculty investigators and several biopharmaceutical researchers.

- **Molecular Libraries Probe Production Centers Network (MLPCN)[22]:** This NIH NCATS initiative collaborates with researchers from all over the world and advises them on assay design and development, chemistry research, informatics research, technology development projects, and running high-throughput screens and chemically optimizing small-molecule leads. MLPCN offers access to thousands of small molecules—chemical compounds that can be used as tools to probe basic biology and advance our understanding of disease. Small molecules can help researchers understand the intricacies of a biological pathway or form starting points for novel therapeutics. One of the members of the MLPCN is The Broad Institute's Probe Development Center (BIPDeC). It offers access to a growing library of over 330,000 compounds for large-scale screening and medicinal chemistry. BIPDeC is open to researchers at the Broad and beyond.[23] MLPCPN has collaborated with a large number of organizations such as U.S. Environmental Protection Agency (EPA), the National Toxicology Program (NTP), National Institute of Environmental Health Sciences (NIEHS),

FDA, National Cancer Institute (NCI), many rare disease foundations, and other intramural and extramural laboratories. Its purpose is to perform a large number of high-throughput screens on molecular targets and cellular phenotypes relevant to almost every area within biology and disease.[24]

- **The EU-OpenScreen[25]:** This is a European Research Infrastructure Consortium (ERIC) founded in April 2018 by seven European Countries: the Czech Republic, Finland, Germany, Latvia, Norway, Poland, and Spain, with Denmark joining as a full member in 2019, while some other European countries are preparing to join. This initiative is the European Counterpart of the US NIH Roadmap initiative. It integrates high-capacity screening platforms throughout different European countries. Users can have access to a rationally selected compound collection, comprising up to 140.000 commercial and proprietary compounds collected from European chemists. This consortium encourages the participation of and offers open access to its shared resources to academic institutions, small- and medium-size enterprises (SMEs), and industrial organizations.
- **The International Genomics Consortium (IGC)[26]:** Based in Arizona, IGC is a nonprofit, medical research organization that supports and accelerates the translation of genomic discoveries into medically and scientifically useful information for physicians, researchers, and industry. The purpose of IGC is to apply this knowledge in medical management and validate findings that will have an impact in the creation of novel medical breakthroughs. This consortium includes the The Cancer Genome Atlas (TCGA) project of the NIH, which helps provide the genomic "blueprints" of specific cancer types for use by the cancer research community; and the Expression Project for Oncology (expO), which is a national biospecimen repository that is available to assist nonprofit and for-profit research worldwide. The overall goal of the IGC is to acquire data for diagnostics, treatment, and prevention of cancer and complex diseases by molecular profiling, and characterization of differential biomarker expression for personalized medicine.
- **The Novartis Foundation[27]:** This important philanthropic organization is applying the Core Model to help the health care situation of some of the poorest parts of the globe via partnership with multiple organizations. This foundation's objective is "to have a transformational and sustainable impact on the health of low-income communities ...

through a mix of programmatic work, health outcomes research, and their translation into policy." (See Table 7.5 for a list of its partners.)

Though the examples provided here are only small samples of all the initiatives taking place all over the world in the same direction, they speak of a major shift in the way in which the complex process of drug discovery

Table 7.5 Nonexhaustive list of Novartis Foundation partners.

Government organizations
- Department of Health, Philippines
- Ghana Health Service, Ghana
- Ministry of Health, Ghana
- Ministry of Health and Social Services, Namibia
- Ministry of Health and Social Welfare, Tanzania
- Ministry of Health, Burkina Faso
- Ulaanbaatar City Municipal Government and Health Department, Mongolia
- Mongolian National Chamber of Commerce, Mongolia
- Government of South Africa, South Africa

Academic organizations
- École polytechnique fédérale de Lausanne
- Erasmus MC
- Ghana School of Public Health
- Hanoi School of Public Health
- London School of Hygiene and Tropical Medicine
- Malaria Elimination Initiative of the Global Health Group at University of California, San Francisco (UCSF)
- Multidisciplinary Research Center at University of Namibia (UNAM)
- National Academy of Sciences (Health and Medicines Division)
- Nelson Mandela University
- Philippine Council for Health Research and Development
- Swiss Tropical and Public Health Institute (Swiss TPH)
- University of Basel

Nongovernment organizations
- Clinton Health Access Initiative (CHAI)
- Family Health International (FHI) 360
- Foundation Terre des Hommes
- International Federation of Anti-Leprosy Associations (ILEP) members:
  - American Leprosy Missions (ALM)
  - German Leprosy and Tuberculosis Relief Association/ German Leper Relief Organization (GLRA/DAHW)
  - Netherlands Leprosy Relief (NLR)
  - FAIRMED (former 'Leprosy Relief Emmaus Switzerland')

Continued

Table 7.5 Nonexhaustive list of Novartis Foundation partners.—cont'd

- Program for Appropriate Technology in Health (PATH)
- Tanzanian Training Center for International Health (TTCIH)
- The International Consortium for Health Outcomes Measurement (ICHOM)
- Onom Foundation
- IntraHealth International
- Centers for Disease Control and Prevention (CDC) Foundation
- American Heart Association

International organizations

- United Nations Children's Fund (UNICEF)
- World Health Organization (WHO)
- National Council on Disability (NCD) Alliance

Private sector

- Dimagi Inc
- Metahelix, India
- Novartis Philippines
- VOTO, Ghana
- Intel

Memberships

- Consultative status at the United Nations Economic and Social Council (UN ECOSOC)
- Member, Medicus Mundi Switzerland Network
- Member, Board of the Tanzanian Training Center for International Health
- Member, Broadband Commission for Sustainable Development
- Member, Global Health Group Advisory Board, Global Health Group, University of California, San Francisco
- Member, Advisory Group for Organization for Economic Cooperation and Development (OECD) Development Center Network of Foundations Working for Development (netFWD)
- Member, Commission for Research Partnerships with Developing Countries (KFPE)
- Member, Neglected Tropical Diseases (NTD) Modeling Consortium
- Member, Governing Council UN Technology Bank for the Least Developed Countries
- Member, European Foundation Center (EFC)
- Member, Global Partnership for Zero Leprosy
- Forum Member, Institute of Medicine (IOM) Forum on Public—Private Partnerships for Global Health and Safety (PPP Forum)

The Novartis Foundation convenes and works with local and global partners to catalyze sustainable health care models.
From Novartis Foundation website; https://www.novartisfoundation.org/about-us/our-partners *(original source).*

and development is being tackled by many different organizations, including the biopharmaceutical industry, in the entire world. More importantly, they are proof of the validity of the Core Model and its current application. All of these initiatives could have a great impact in saving time, money, and human effort, allow the development of safer and more effective medicines, and have a great impact in making health care more accessible.

Innovation deficit?

As I said in Chapter 3, most pharmaceutical executives and even academics usually complain that the pharmaceutical industry's major problem is lack of innovation. Given all the examples already provided throughout this book, it is clear that this is not the case at all. If anything, this is an unprecedented time in history, one in which we actually have an *excess* of innovation. Moreover, it has become very difficult to find ways to make all this knowledge converge so that somehow it begins to make sense and becomes useful to society in the form of better technological tools and, specifically, better drug products. Therefore, the need exists to pay attention to the managing of knowledge and leadership.

I believe that scientific specialization has made it more difficult for scientists to integrate effectively the scientific knowledge that is generated in the world on a daily basis. There is simply too much information to be processed and fully integrated into a coherent whole. The knowledge and new technology amassed in past decades has been growing exponentially, yet the conversion of that knowledge into fundamental scientific paradigms that will explain the behavior of cells, tissues, organs, and systems, and then the use of this holistic understanding in the creation of concrete commercial products has lagged. This is the root of all the problems in R&D productivity that the industry has at present and the reason why no health care system will work, no matter how much money is infused into reforming it, unless this issue is fully addressed.

Endnotes

1. Sánchez-Serrano, I., 2011. The World's Health Care Crisis: From the Laboratory Bench to the Patient's Bedside. Elsevier.
2. See endnote 1.

3. Office of Technology Development, March 21, 2016. Collaboration to develop cancer therapeutics. The Harvard Gazette. https://news.harvard.edu/gazette/story/2016/03/collaboration-to-develop-cancer-therapeutics/.
4. See Columbia University Institute for Genomic Medicine's website: http://www.igm.columbia.edu/collaborations/industry-partnerships.
5. See AstraZeneca's website:https://www.astrazeneca.com/media-centre/articles/2017/harnessing-the-power-of-genomics-through-global-collaborations-and-scientific-innovation-12012018.html.
6. See Pfizer's website: https://www.pfizercti.com/about_cti.
7. Zastrow, M., December 8, 2017. The top academic and corporate partners in the Nature Index. Nature. https://www.natureindex.com/news-blog/the-top-academic-and-corporate-partners-in-the-nature-index.
8. See Novartis' website: https://www.novartis.com/news/media-releases/novartis-expands-global-collaboration-amgen-commercialize-first-class-amg-334.
9. See Novartis' website: https://www.novartis.com/news/media-releases/novartis-announces-clinical-collaboration-pfizer-advance-treatment-nash.
10. See Eli Lilly's: https://openinnovation.lilly.com/dd/includes/pdf/OIDD_Brochure.pdf.
11. See Eli Lilly's: https://openinnovation.lilly.com/dd/.
12. See Eli Lilly's website: https://investor.lilly.com/news-releases/news-release-details/eli-lilly-and-company-announces-new-drug-discovery-initiative.
13. See Bayer's website: https://innovate.bayer.com/.
14. BusinessWire, January 25, 2016. Apollo Therapeutics: Consortium of World-Leading UK Universities and Global Pharmaceutical Companies Launch £40 Million Fund to Drive Therapeutic Innovation, BusinessWire. https://www.businesswire.com/news/home/20160124005065/en/Apollo-Therapeutics-Consortium-World-Leading-UK-Universities-Global.
15. Chaguturu, R., et al., 2014. Collaborative Innovation in Drug Discovery. Wiley.
16. Walker, N., December 7, 2017. Accelerating Drug Development Through Repurposing, Repositioning and Rescue. PharmaOutsourcing. https://www.pharmoutsourcing.com/Featured-Articles/345076-Accelerating-Drug-Development-Through-Repurposing-Repositioning-and-Rescue/.
17. Brown, A.S., Patel, C.J., 2017. A standard database for drug repositioning. Scientific Data 4, article number: 170029.
18. See NCATS website: https://ncats.nih.gov/ctsa/about.
19. See endnote 15.
20. Marusina, K., et al., Winter 2011. The CTSA Pharmaceutical Assets Portal — a public—private partnership model for drug repositioning. Drug Discovery Today: Therapeutic Strategies 8(3—4), 77—83. https://www.ncbi.nlm.nih.gov/pmc/articles/PMC3388510/.
21. See NCATS′ website: https://ncats.nih.gov/pubs/features/ctsa-collaboration.
22. See NIH's website: https://commonfund.nih.gov/molecularlibraries/index.
23. See Broad Institute's website: https://www.broadinstitute.org/mlpcn/mlpcn-small-molecules-big-impact.
24. http://grantome.com/grant/NIH/ZIB-TR000004-03.
25. See EU-Opescreen website: https://www.eu-openscreen.eu/.
26. See the International Genomics Consortium's website: http://www.intgen.org/.
27. See Novartis Foundation's website: https://www.novartisfoundation.org/.

CHAPTER 8

The Core Model, innovation, economic growth, development, and policy-making

Throughout this book, we have presented the Core Model and illustrated its effective use in drug discovery and development through the specific case story of the development of bortezomib. We have explored it in the context of innovation and leadership, and have discussed how it could solve the "Tragedy of the Anticommons" created by Intellectual Property Rights. We have also seen how it is gradually becoming implemented by the bio-pharmaceutical industry, academia, the federal government, and nonprofit organizations to save time, money, and labor. The Core Model can increase knowledge that will eventually become innovation, invention, and applied science; and, finally, we have highlighted its great potential in global health care if implemented in a deliberate and conscientious manner.

One of the important revelations of this paradigm, at least in my opinion, is how it can account for the generation, translation, and expansion of innovation via the bidirectional collaboration between the Core, the Bridge, and the Periphery. This is, to my knowledge, the first time that the origin and application of innovation is explained in a systematic manner. The constant feedback-loop mechanism between these three constituents creates and fosters innovation, which in the end promotes economic development and growth. Innovation always starts at the Core level (*This statement takes the term Core in its absolute conception; for instance, peripheral entities, such as the NIH, can adopt, as we have seen, a Core modality.*) with the goal of satisfying Peripheral (societal) demands. Via the Bridge, innovation becomes enriched and propagated until it finds the desired and even unexpected use in society. This model works both at the microeconomic and macroeconomic levels. Moreover, it can work for any human enterprise.

Having an understanding of this process can be very useful to firms as well as to society (through its policy agencies) in terms of how to develop strategies for the development of innovation and how to translate in an

The Core Model
ISBN 978-0-12-814293-6
https://doi.org/10.1016/B978-0-12-814293-6.00008-1

intelligent manner that innovation without having to re-invent the wheel, while contributing to economic growth and development. Therefore, next I list a series of initiatives at the public policy level and also at the pharmaceutical level that can help the system.

Academia

Although collaboration among academic scientists from all over the world has always been a natural activity in this profession, collaboration with industrial partners has not always been as smooth—most notably since the creation of the Bayh–Dole Act of 1980.

Back in 2006, when I first published my article on the story case of the development of bortezomib (written in 2004) and *The World's Health Care Crisis* (published in 2011), the lack of understanding between academia and industry was very significant and had a big weight on the attitude and, perhaps, even the outcomes of collaboration. Over the years, thanks to some successful case-story collaborations between academia and industry, the collaboration between academia and industry, at least in the United States, has become more fluid. Reasons for this included the advent of translational research, the multiple initiatives launched by the U.S. Government, such as the NIH Roadmap and the FDA Critical Path Initiative (among others), the sequencing of the human genome, the fabulous advances that have taken place in bioinformatics, etc.

One of the major problems in basic academic research, as I have pointed out before, is that this activity is greatly driven and influenced by personal ego. In addition, oftentimes, ego becomes vicious, unhealthy, and accompanied by extreme selfishness, ruthlessness, inconsideration, divisiveness, psychological abuse, dishonesty (intellectual and otherwise), and other destructive behaviors that are not very conducive to good collaborative working dynamics. Because scientists in academic settings need to constantly prove themselves via groundbreaking discoveries or models so that they could be "the first" to publish in *Nature* or *Science* magazines (or other high-impact journals), but also to be able to secure grants, tenure, and collegial–societal prestige and recognition, a culture of "secrecy" and "closeness" (and sometimes even paranoia) typifies some academic research and training.

There are, in fact, three types of academic researchers: first, the ones who are not interested in collaborations with industry. Second, the ones who are open to the possibility of working with industry and do so when the opportunity arises; however, they always remain focused on basic academic research. Third, the ones who are very open to work with industry, and engage in such collaborations all the time, may even leave academia to work in industry. Therefore, it is very important that this differentiation is taught at the university level to undergraduate and graduate students in the life sciences, as well to medical students. Though basic research should never be neglected, it is important that scientists are taught and encouraged to translate their basic discoveries into biomedical applications, communicate those findings to their technology transfer offices, and be open to collaborate with industry. Courses and workshops on how to accomplish this should be provided in academic settings (and universities such as Harvard and Massachusetts Institute of Technology (MIT), as I have been able to witness, are excellent at this), as they will create a better understanding of why this is important.

Public policy funding

In the United States, the major single investor for basic, applied, and translational research is the National Institutes of Health. The NIH budget for Fiscal Year (FY) 2018 was US$ 37 billion and it received a US$ 2 billion boost for 2019.[1]

However, according to some data[2] in the year 2015, for the first time since World War II, the federal government no longer sponsors the majority of basic research in the United States. Surveys from the National Science Foundation (NSF) reveal that federal agencies provided only 44% of the total $86 billion spent on basic research in 2015. The federal share, which topped 70% throughout the 1960s and 1970s, stood at 61% as recently as 2004 before falling below 50% in 2013 (see Graph 8.1). This has been attributed to two main reasons: first, the flattening of US federal investment in basic research; and, second, an increase in the corporate side of funding into this category. The major driver in the recent increase in investment in basic research coming from the corporate side has been, not surprisingly, the pharmaceutical industry.

In fact, the pharmaceutical investment in basic research has seen a dramatic increase over the last decade. In 2008, its investment in this sector was $3 billion; in 2014, it soared to $8.1 billion in2014.[3] On the other hand we need to consider that, even though the NIH has received a funding increase in the last couple of years, from FY 2003 to 2015, it lost 22% of its capacity

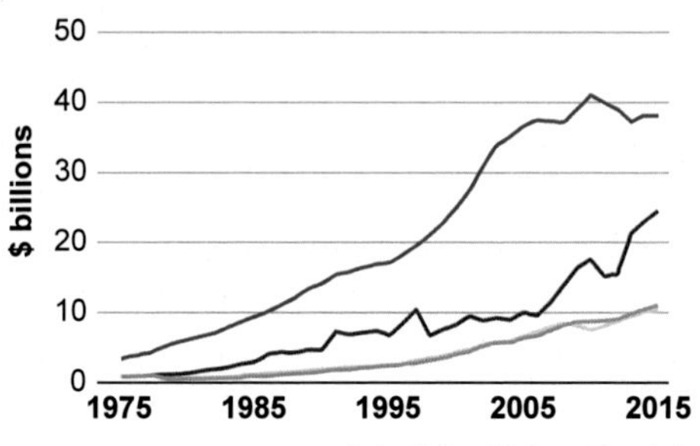

Graph 8.1 Investment in basic research from 1975 to 2015. *(Source: Science. http://www.sciencemag.org/news/2017/03/data-check-us-government-share-basic-research-funding-falls-below-50. (Reprinted with permission.))*

to fund research due to budget cuts, sequestration, and inflationary losses. Its economic situation still remains suboptimal (see Graphs 8.2A–C).[4]

Given all that has been exposed in this book, it is somewhat pre-occupying that Big Pharma drives the game, because it is the Periphery's role to provide the necessary funding for the benefit of society. If the

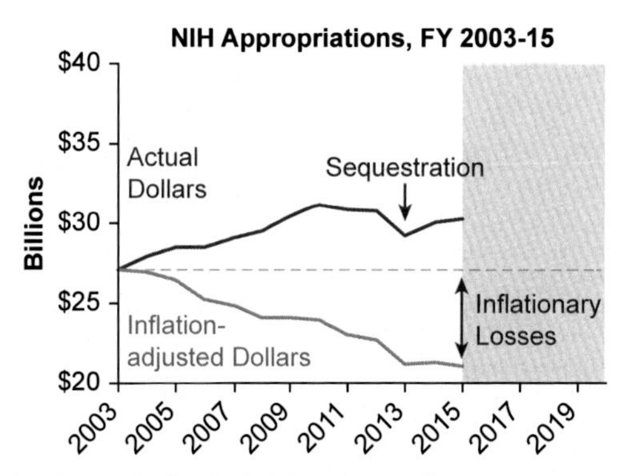

Graph 8.2A ***NIH funding trends. Federal support of NIH.*** U.S. Biological and Medical Research Fell for Over a Decade From FY 2003 to 2015, the National Institutes of Health (NIH) lost 22% of its capacity to fund research due to budget cuts, sequestration, and inflationary losses. Reduced funding capacity results in fewer grants, fewer new discoveries, and talented scientists leaving research.

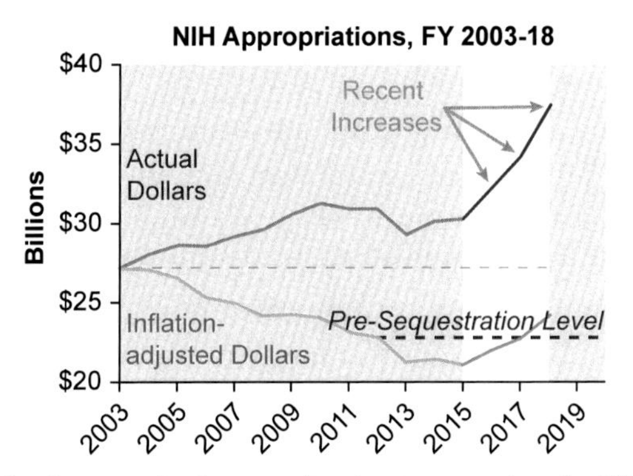

Graph 8.2B ***NIH funding trends. Congress has begun restoring the NIH budget. For the last 3 years (FY 2016, 2017, and 2018) Congress raised the NIH budget.*** With the FY 2017 budget increase alone, NIH was able to award 1149 more research project grants than had been projected. These budget increases reversed losses stemming from sequestration.

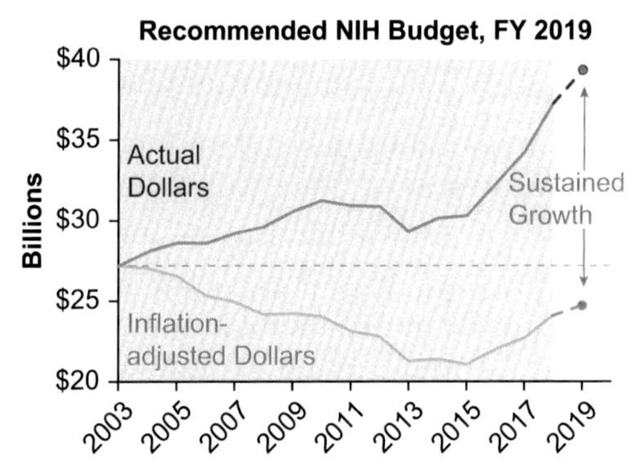

Graph 8.2C ***NIH funding trends. We must sustain this progress. Federation of American Societies for Experimental Biology (FASEB) recommends at least $39.3 billion for NIH in FY 2019.*** Additional funding will allow NIH to support even more meritorious research to improve the country's health and quality of life. *(Source: Federation of American Societies for Experimental Biology (FASEB), based on National Institutes of Health (NIH) data; http://www.faseb.org/Portals/2/PDFs/opa/2018/Factsheet%20Restore%20NIH%20Funding.pdf. (Reprinted with permission.))*

corporate sector rules research, it will then rule health care. There is actually the potential danger that even if the process of developing novel drugs becomes optimized, then pharmaceutical companies will continue to abuse both the IPRs as well as the pricing of new medications. See Graph 8.2 on the amount of investment provided by the NIH since 2003.

Although the NIH through its many institutes, such as the National Cancer Institute, and divisions, such as NCATS, it is doing a terrific job being very creative at fostering and implementing collaboration, in a Core Model fashion. The NIH accelerates drug discovery and its translation into medical products, and thus requires much more funding, not only to sustain the initiatives already in place until they bear the expected fruits, but also to create new initiatives. These could include new consortia and sweeping initiatives such as the creation of a map of the biological sex differences between women and men, among others that will have a big impact in precision and personalized medicine.

A very helpful initiative that the NIH or the federal government could create is the setting up of public Contract Research- types- of Organizations for start-ups and more grants to cover for the licenses fees associated with the translation on basic research into commercial products. This is an extremely important point, because drug discovery and development has many barriers to entry. A start-up requires at least US $5—10 million dollars from the licensing point to the development of a clinical candidate, and, in most instances, the technologies are not quite ready to secure venture capital funding, thus affecting the potential number of drugs that could eventually be approved. We need to create a "Poor Man's Mechanism to Drug Discovery and Development." Other mechanisms such as the provision of "small" (e.g., rodent) and "large" (e.g., dog, pig, nonhuman primates) animal models for rare diseases should be developed, as well as more grants for start-ups to cover some of the expenses to get the company going.

In addition, the federal government should create a mechanism to foster more interdisciplinary collaboration. So, in the same way in which scientists have been rewarded in the past for their original research and for their entrepreneurship, they should also be rewarded for their collaborative work. Progress in this area will certainly depend on a positive attitude in this front.

On the other hand, the FDA is another agency that needs more funding and is understaffed. As I have suggested before, the FDA should be broken down into two separate agencies, one focused on food and the other focused on medical products, like other mature regulatory agencies of the same kind in the world.

Pharmaceutical industry level

With the very large number of mergers and acquisitions that have occurred over the last two decades, pharmaceutical companies have become

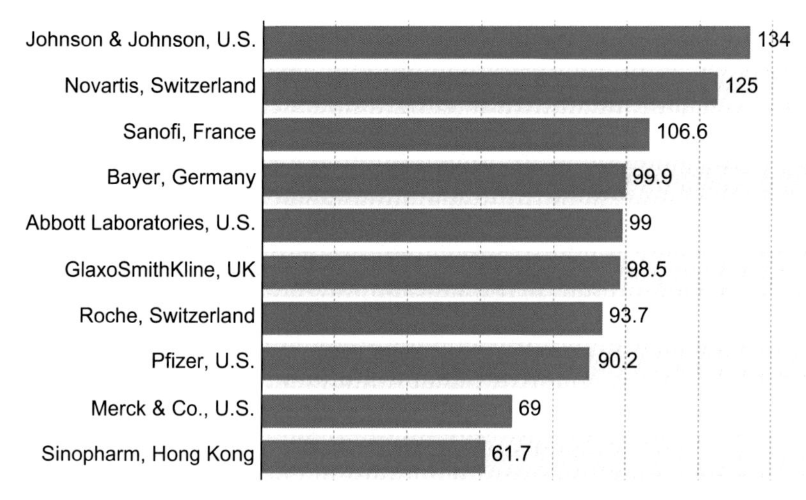

Graph 8.3 ***2018 Ranking of the global top 10 biotech and pharmaceutical companies based on employee number (in 1000).*** *(Source: Statista and other Internet sources. https://www.statista.com/statistics/448573/top-global-biotech-and-pharmaceutical-companies-employee-number/. Reprinted with permission.)*

extremely large. See, for example, Graph 8.3 for a list of the top 10 pharmaceutical companies in terms of the numbers of employees worldwide. With such a large number of people, it becomes very difficult to lead a firm, not to mention a productive firm. Sometimes these companies have more than 20 layers of management. No company can be very efficient in that way. Thus, breakdown of their facilities and divisions is unavoidable, which is what is really happening with companies such as GlaxoSmithKline, Johnson & Johnson, Pfizer, AstraZeneca, Eli Lilly, Sanofi, Merck, Novartis, etc. For this reason, it is very important that the pharmaceutical industry become even more open to external collaboration in a Core Model fashion and continue to create the type of agreements that it has already created with the NIH and academic centers in different parts of the world, as described in the previous chapter. The creation of "standards" as has happened in other industries can benefit research at the precompetitive level.

On the other hand, biotech start-ups have been adopting the Core Model by going "virtual"—that is, companies with no labs, in which a great deal of the preclinical and clinical studies are outsourced to external Contract Research and Medical Organizations for a fee. This strategy has become, in fact, a new trend, because it simplifies the drug discovery and development process, given that a large number of assays such as absorption, distribution, metabolism, and excretion/toxicity (ADME/tox), etc., can be outsourced to specialized firms.

Private foundations

On top of all the activities which have already been discussed throughout this book, private foundations and philanthropic organizations can be instrumental in the creation of pharmaceutical markets and sponsoring drug development initiatives, as was the case of the Cystic Fibrosis Foundation and the drug Kalydeco. Financing early stage firms by helping them with licensing fees, preclinical studies, and early stages of clinical trials can certainly increase the changes of bringing to the market new drugs, in particular in areas such as orphan and neglected diseases. Another way in which private foundations can help is creating Clinical Research Organizations (CROs) and/or financing and the creation of animal models for their specific disease cause. About 8000 rare diseases exist and counting; therefore, a great need exists for such platforms as well as more collaboration and cross-communication among disease foundations to create these platforms and to avoid repletion of efforts. One of the great difficulties for entrepreneurs and early stage biotech companies is to have the adequate animal models, especially in the field of genetic disorders, because it take a great deal of time, investment, and expertise to develop (and breed) such models, in particular in "large" (nonhuman primates) animals. Without adequate, nonrodent animal models to test their compounds before arriving to phase I clinical trials, small biotech companies suffocate. This is very important. Another way to help start-ups is by recruiting patient populations in their particular disease cause.

The developing world

Besides its implementation and importance at all stages of drug discovery and development in the industrialized world, as we have seen throughout the book, the Core Model is ideal for use in the developing world as well. Perhaps in the developing world due to unthreatening pharmaceutical competition and the evolving nature of their research, pharmaceutical, and health care systems it would be easier for the governments to foster collaboration within all sectors of the "Core Model."

For instance, the Cuban government has been very closely involved in the creation and development of a strong biotechnology industry. It has been designed as an element of the state-funded health care system with the main objective of creating medicines to satisfy the national market's medical needs (of the 857 medicines on Cuba's list of medicines approved for use in the national health system, 569 are produced domestically).[5] But the role of the government has encompassed not only investment, but also strategic planning, a research and development and production infrastructure, as well as the training of a qualified workforce. This has resulted in the creation of many different, but highly networked research institutes including the West Havana Biocluster, within which Cuba's Center for Genetic Engineering and Biotechnology (Centro de Ingeniería Genética y Biotecnología, CIGB).[6] CIGB is one of the most important companies in Cuba with an impressive record of producing innovative biotech products for the country's health care system. In 2012, the Cuban government established BioCubaFarma, the state umbrella agency overseeing the 31 state-run pharmaceutical and biotechnology R&D and production facilities. As of 2013, this agency had more than 21,000 employees including 6158 university graduates (270 with a Ph.D. and over 1000 with a master's degree).[7] Interdisciplinary collaboration and knowledge sharing are at the center of Cuba's biotechnology strategy as well as long-term state-fueled integration of a multiinstitutional system, including academia, hospitals, and teaching centers. These include the National Coordinating Center of Clinical Trials (Centro Nacional Coordinador de Ensayos Clínicos [CENCEC])[8] and the health care system through its Center for State Control of the Quality of Medicines (Centro para el Control Estatal de Medicamentos, Equipos, y Dispositivos Médicos [CECMED]).[9] All these characteristics may actually account for the high innovation rate and success achieved by the Cuban biotechnology industry.

When it comes to Cuba's nationally produced intellectual property, the patents of the Cuban industry are owned by the government agency, which avoids the "Tragedy of the Anticommons." This agency functions as a kind of patent pool, in which every firm has the possibility of using complementary knowledge to advance new products. In practice, companies cooperate with each other informally. This resembles more an internal open source of innovation, which is no surprise at all, as the notion of "cooperation instead of competition" is one of the most promoted values in the industry at a national level.

As a result, Cuba has launched a large number of biotech products that are the envy of many industrialized countries' biotech and pharmaceutical companies.[10,11] Therefore, in a fully integrated Core Model, a small and isolated country such as Cuba can provide the medicine necessary for its citizens. Despite the end of former Soviet Union subsidies in 1991 and imposition of the US embargo against Cuba, this country has one of the highest life expectancy rates in the region, with the average citizen living to 78.8 year old (76.5 males; 81.3 females)[12] higher than one of the United States: 78.6 (81.1 for females; 76.2 for males).[13]

Can we imagine how other countries in the world would benefit from the Core Model?

Endnotes

1. Kaiser, J., September 14, 2018. NIH gets $2 billion boost in final 2019 spending bill. Science. http://www.sciencemag.org/news/2018/09/nih-gets-2-billion-boost-final-2019-spending-bill.
2. Mervis, J., March 9, 2017. Data check: U.S. government share of basic research funding falls below 50%. Science. http://www.sciencemag.org/news/2017/03/data-check-us-government-share-basic-research-funding-falls-below-50.
3. See endnote 2; according to NSF's annual Business Research and Development and Innovation Survey (BRDIS), which tracks the research activities of 46,000 companies.
4. FASEB, 2018. NIH Research Funding Trends. http://faseb.org/Science-Policy–Advocacy-and-Communications/Federal-Funding-Data/NIH-Research-Funding-Trends.aspx.
5. Notman, N., March 16, 2018. Cuba's cancer treatments. Chemistry World. https://www.chemistryworld.com/features/cuba-socialism-cigars-and-biotech/3008585.article.
6. See the CIGB's website: https://www.ecured.cu/Centro_de_Ingenier%C3%ADa_Gen%C3%A9tica_y_Biotecnolog%C3%ADa; accessed 3/9/2019
7. Notman, N., See endnote 6.
8. See CENCED's website: http://instituciones.sld.cu/cencec/centro-nacional-coordinador-de-ensayos-clinicos-cencec/.
9. See CECMED's website: https://www.cecmed.cu/.
10. Notman, N., See endnote 9.
11. Medical Daily Journal, December 20, 2017. Cuba exports medicine to dozens of countries. It would like the U.S. to be one of them. Medical Daily Journal. https://medicaldailyjournal.com/2017/12/20/cuba-exports-medicine-dozens-countries-like-u-s-one/.
12. See Index Mundi. https://www.indexmundi.com/cuba/life_expectancy_at_birth.html.
13. Scuti, S., November 29, 2018. Drug overdoses, suicides cause drop in 2017 US life expectancy; CDC director calls it a wakeup call, CNN. https://edition.cnn.com/2018/11/29/health/life-expectancy-2017-cdc/index.html.

PART IV

CHAPTER 9

Conclusion: the Core Model: a novel economic theory

I began investigating the multiple factors that could account for the success and failure of biotechnology firms in the United States and Europe in 2003–04. Some of the variables involved in that analysis—eventually published in 2006 in *Nature Reviews Drug Discovery*[1]—were: (1) the relationship between early stage biotechnology firms and their parental academic organizations and other academic institutions; (2) the collaborations that these firms established with pharmaceutical companies and private investors; (3) the role of federal-funding agencies in the translational process between academia–industry/private–public sector; (4) the support that early stage firms received from philanthropic organizations and advocacy groups; and (5) considerations of the impact of secrecy and intellectual property (IP), as described by the "Tragedy of the Anticommons", a type of paradoxical setting, in which too many separate rights-holders of a single resource can block each other's use of that resource.

I have studied the story case of many successful and failed biotechnology companies throughout the world, including the story case of the development of bortezomib for the treatment of multiple myeloma by the Cambridge (US)-based biotechnology company Myogenics/ProScript (later acquired by Millennium Pharmaceuticals). This one case seemed to me both highly unusual and a quintessential example of what it takes to develop a drug in the biotechnological field.

Careful study of the story that I had built on how Myogenics/ProScript developed bortezomib revealed that, despite numerous challenges, difficulties, and lack of funding, this company deduced a "Proof of Principle" in animal and human models very quickly and with very few resources. This was in stark contrast with what generally happens in industry. Further examination of the case revealed that even though boronates were an innovative class of drugs—initially developed as serine protease inhibitors by Dupont/Merck—the fact that the original compound failed in Phase II clinical trials as a treatment for emphysema, gave them a bad reputation

The Core Model
ISBN 978-0-12-814293-6
https://doi.org/10.1016/B978-0-12-814293-6.00009-3

among medicinal chemists. Therefore, the industry was very wary and skeptical of bortezomib, being a molecule belonging to this class—not to mention that it was considered too toxic. Another great innovative idea was to try to inhibit the newly characterized organelle, the proteasome, with a series of synthesized compounds, including, among several others, bortezomib. However, both initiatives, independently of all the other obstacles, such as lack of sufficient funding, etc., and regardless of its high level of innovation, were doomed to perish if it were not because the company made the following decisions. First, it decided to test the initial inhibitors via collaboration in the labs of the founders of the company. Second, it decided to give the compound freely to other academic scientists from different parts of the world to test the effects of proteasome inhibitors in vivo and in vitro. Third, it decided to establish multiple collaborations with academic groups, teaching hospitals, pharmaceutical companies, federal agencies such as NIH/NCI, and advocacy groups at the same time. These interactions were performed out of necessity (the hard reality of lack of funding and resources, including labor, materials, instrumentation, facilities, etc.) and despair (at which point the only thing that could save is "common sense").

These interactions were possible thanks to the initiative and audacity of the company's leaders (including its founders) and the championing of individuals such as Julian Adams and the commitment and talent of other members of the company such as Vito Palombella, Peter Elliot, Ross Stein, et al.—all of whom capitalized on their personal connections and previous training to secure collaborations with outside groups. These individuals brought to the company key figures as Scientific Advisory Board Members, who, in turn, had even more important and heavyweight connections. These new connections became interested in bortezomib and brought to the company essential and extremely valuable knowledge to move forward the bortezomib project, as well as funding, resources, free advice, animal models, and even public awareness through the multiple myeloma and other foundations.

It was thanks to multiple collaborations that a better understanding of how in blocking the proteasome's catalytic active site, bortezomib inhibits an important cellular mechanism that regulates the cell cycle through the activation of nuclear factor-κB (NF-κB), which led to the discovery that preventing NF-κB activation leads to apoptosis and renders malignant cells more vulnerable to chemotherapy and radiation. In addition, these collaborations made it possible to open up the field of potential indications for the drug, specifically to cancer, even though that was not the original plan. In a personal conversation with Julian Adams and Vito Palombella, they told me that the results of the multiple myeloma trial were the sixth attempt at the drug!

Therefore, bortezomib would have not made it to the market had it not been for these collaborations. However, here the company took a most unusual step: without its members knowing what they were doing: transferring knowledge, integrating and assimilating external knowledge, and finally translating the compounded knowledge (internal and external) into a specific application while opening the doors to many other doors. This was a bidirectional interaction in which collaborators—who were people interested in similar or related problems—were doing the same. Both the company and the external collaborators were very conscious of their mission in advancing human knowledge about the behavior of the cell and about the biology of disease and that in finding application of their knowledge they could provide society with economic as well as health care benefits. Unconsciously, however, they (the company and its collaborators) sought help from those structures that society has created (federal-funding agencies, advocacy groups, hospitals, etc.) for the benefit of all. Nevertheless, unlike many other biotechnology companies, Myogenics/ProScript did this exceptionally well. Bortezomib was approved in record time, under fast-track application, using less money and taking less time, overall, than the one usually quoted in industry.[2]

At midnight of April 23, 2004, when under great pressure, while pondering at my computer how bortezomib managed to get to the market at all that all the relationships that I just described in the previous paragraph clearly came to my mind one after the other, like a stream of crystalline water. Suddenly, Fig. 9.1A–C appeared one after the other in my mind's eye, like a series of images projected on a big screen by a slide projector. I realized that I had discovered a paradigm for the way in which bortezomib was developed—a paradigm that could be both conceptualized and proven to actually exist as an organizational model in real practice. Epitomized by the development of bortezomib, the paradigm was illustrated by the other examples provided in this book, such as the way in which many biotech companies are working now "virtually." I called this the Core Model because it seemed to me so basic that it could be applicable to any other human activity.

Therefore, in trying to understand what made biotech companies successes or failures, I never imagined that I would end up discovering what I now consider a new economic model or even a theory, said with

(A)

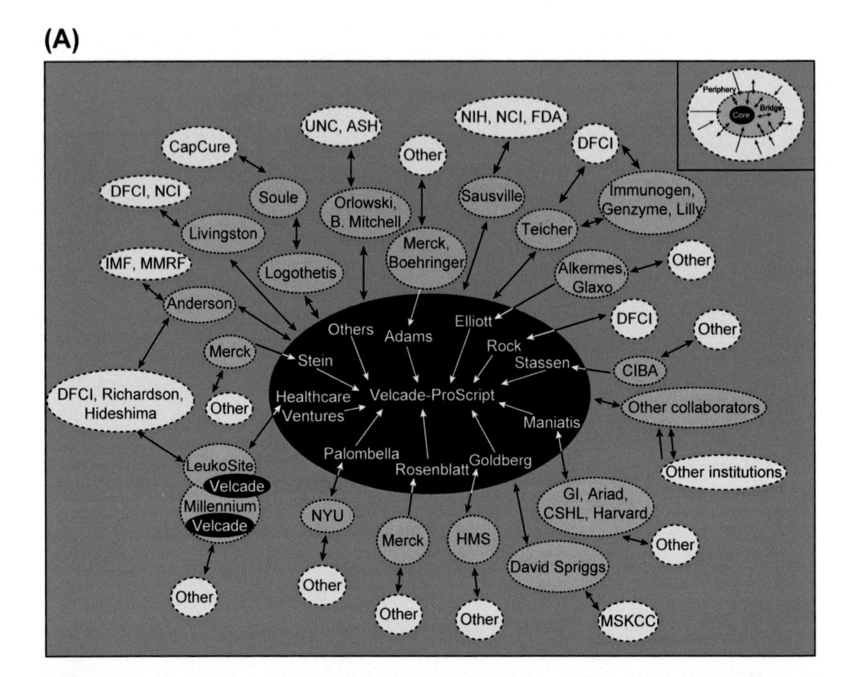

(B)

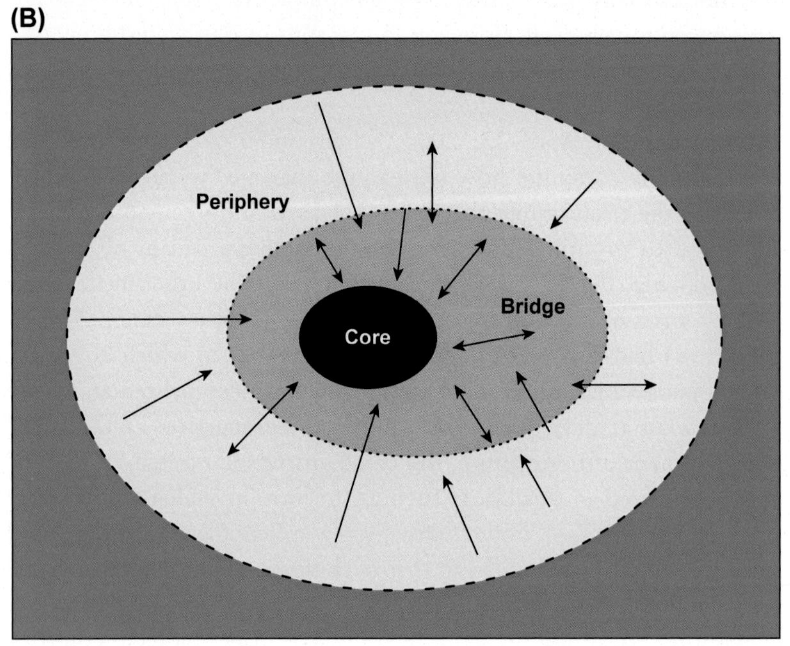

Figure 9.1 (A) Selected collaborations involved in the development of bortezomib, (B) Conceptualization of the interactions in the development of bortezomib as the Core Model, (C) Three-dimensional view of the Core Model. **Key:** Gray = Core; Red = Bridge; Orange = Periphery. All these sectors function through fluid bidirectional interactions.

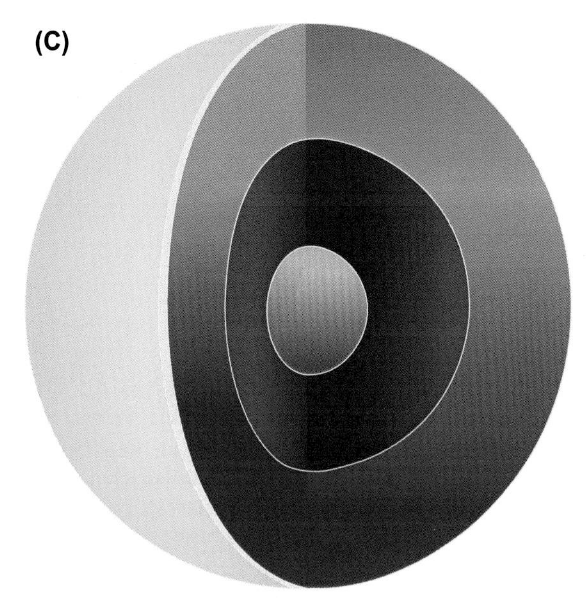

Figure 9.1 Cont'd

humility. It says that productivity (and progress and economic growth) at the micro and macro levels is a direct result of collaboration (that is, exchange or trade of assets) between the Core, the Bridge, and the Periphery, using knowledge transfer, integration, and knowledge translation modus operandi.

This paradigm is valid for any human activity ranging from family, to friendship, to education, large corporations, governments, etc. Moreover, it can be extremely flexible too. It can be used for profit as well as nonprofit activities and organizations. Importantly, it shows that Intellectual Property Rights are not really an inescapable necessity to discover and develop high-quality pharmaceutical drugs. Though, ideally, some type of protection is necessary so that dishonest external players do not milk or steal for selfish financial gains the work that others have done with an altruistic and higher moral intention. In addition, this possibility may need some type of regulatory consideration by the World Intellectual Property Organization, in case it is not already in place.

Throughout the book, we have seen how we can use the Core Model for the improvement of drug discovery and development. However, we can use it too for global health care (see Fig. 9.2). If we consider global health care as our Core, and society—patient populations as our Periphery, we will need to use the pharmaceutical industry, intellectual property,

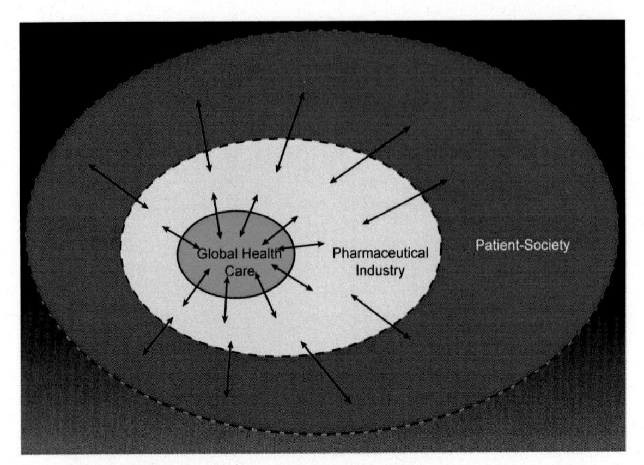

Figure 9.2 The Core Model Applied to Global Health Care. **Key:** Blue = Global Health Care (Core); Yellow = Pharmaceutical Industry, IP, Policy-Making, Regulators, Federal Funding Agencies, Academia, Advocacy Groups, Investors, Philanthropy, Pricing (Bridge); Red = Patient Society (Periphery). Double-headed arrows show bidirectional interactions.

public policies, federal-funding agencies, advocacy groups, investors, not-for-profit organizations, philanthropists, academia, consortia, etc., as our Bridge as a catalyst to lower the activation energy of global health in the service of humankind. This certainly requires self-examination and optimization of the Bridge's components and their realization of the moral obligations that they have with the human race and with the planet.

Finally, as I finish writing this book, two old Latin proverbs used by the ancient Romans come to my mind with a radically new and clear meaning. The first one: "Quid pro quo", "This for that". The second: "Manus manum lavat" ("One hand washes the other"), which later evolved in Italian as: "Una mano lava l'altra e tutt'e due lavano il viso", meaning, "One hand washes the other, and both wash the face". I do not think it would be an exaggeration on my part to believe that the Core Model not only explains how human society works, but also how it started and how it evolved into civilization.

—Ibis Sánchez-Serrano, December 12, 2018. Santiago de Veraguas, Panamá.

Endnotes

1. Sánchez-Serrano, I., 2006.
2. Year 2003.

CHAPTER 10

Epilogue: the Core Model and other industries

This book has demonstrated how the Core Model can be used in the biopharmaceutical industry. However, the Core Model is a universal model that can be used in other industries as well*. In fact, we can see how the model is being used today in both "Closed Innovation" systems (for instance, the aerospace and telecommunication industries, architecture, engineering, automobiles, textiles, arts guilds, and communities, medical device industry, multilevel marketing, etc.), as well as "Open Innovation" Systems (Wikipedia, Wikimedia Commons, Software, some drug-discovery and development consortia, Internet, social networks, mass media, etc.). It can be used in public as well as the private domains in a for-profit manner (in which Intellectual Property [IP] and secrecy lock the system) and in a not-for-profit manner (with or without IP).

In 2014, former Tesla CEO Elon Musk said:

> *Tesla Motors was created to accelerate the advent of sustainable transport. If we clear a path to the creation of compelling electric vehicles, but then lay intellectual property landmines behind us to inhibit others, we are acting in a manner contrary to that goal. Tesla will not initiate patent lawsuits against anyone who, in good faith, wants to use our technology.*[1]

* A recent and very beautiful example of the use of the Core Model in a large scale can be illustrated in the field of astronomy. The Event Horizon Telescope (EHT), a collaborative network of eight radio telescopes spanning locations from Antarctica to Spain and Chile, in an effort involving more than 200 scientists and 100 academic and research institutions from all over the world, published on April 10, 2019, the first image ever of a black hole. This extraordinary and historic feat, which a few years ago seemed something almost impossible, was achieved thanks to the collaboration and exchange of assets between the Core, the Bridge, and the Periphery using knowledge transfer, integration, and translation. For more information see: https://eventhorizontelescope.org/; https://www.youtube.com/watch?v=omz77qrDjsU&t=10s; and: https://www.theguardian.com/science/2019/apr/10/black-hole-picture-captured-for-first-time-in-space-breakthrough; accessed 4/10/2019.

The Core Model
ISBN 978-0-12-814293-6
https://doi.org/10.1016/B978-0-12-814293-6.00010-X

This is an idea that the pharmaceutical industry may want to consider seriously, given that times are changing. As Thomas Lönngren, former head of the European Medicines Agency (EMA), once said to me in London:

One of the hurdles in drug development is the issue of sharing of information. There is so much research now going on that more and more data will be generated, you know, huge amounts of data will be generated in individual pharmaceutical businesses ... small and big ones, and they all want to bury this in their archives in order not to share it with their competitors ... and I don't think the world can afford to put so much money into research when more or less all this research will be hidden in some archives somewhere ... It does not fly in the future.[2]

It is, therefore, time that the pharmaceutical industry as a whole embrace the generation of a large and common pool of information and technologies that could be shared among members as well as with external players to accelerate the translation of basic discoveries into commercial products. This issue should also be part of IP reform in which still important information cannot be used because some specific types of patents block its usage.[3] If IP is an issue, then eBay or Amazon types of models for IP could work.

When it comes to one of the major problems in the pharmaceutical industry, I completely agree with my friend Rathnam Chaguturu when he points out that

none of the pharmaceutical companies are willing to share the reasons for the failure of their clinical candidates in real time to effectively navigate the 'industry' from committing the same mistakes. It is time for the pharmaceutical industry to embrace, metaphorically speaking, a community-driven 'Wikipharma', a Wikipedia- or 'Waze'-type shared knowledge, openly accessible innovation model to harvest data and create a crowd-sourced path toward a safer and faster road to the discovery and development of life-saving medicines. Pharma ought to give serious consideration to such a game-changing concept.[4]

All of this makes sense within the Core Model, so why not explore its potential?

Endnotes

1. Musk, E., June 12, 2014. All our patent are belong to you. Tesla Blog. https://www.tesla.com/blog/all-our-patent-are-belong-you.
2. Interview with Thomas Lönngren, London, UK, April 2007.
3. Sánchez-Serrano, I., 2011. The World's Health Care Crisis. See endnote 2.
4. Sánchez-Serrano, I., Pfeifer, T., Chaguturu, R., 2018. Disruptive approaches to accelerate drug discovery and development. Part I. Tools, technologies and the core model. Drug Discovery World (Spring), 39–52. https://www.ddw-online.com/business/p322101-disruptive-approaches-to-accelerate-drug-discovery-and-development-(part-1).html.

References

Below is a list of selected writings and media that could be of interest for further reading.

Adams, J., 2004. In: Adams, J. (Ed.), Cancer Drug Discovery and Development: Proteasome Inhibitors in Cancer. Humana, Totowa, pp. 17–38.

Bower, J., Christensen, C., 1995. Disruptive technologies: catching the wave. Harvard Business Review (January–February).

Chandler, A.D., 2005. Shaping the Industrial Century: The Remarkable Story of the Evolution of the Modern Chemical and Pharmaceutical Industries. Harvard University Press.

Chaguturu, R. (Ed.), 2014. Collaborative Innovation in Drug Discovery. Wiley, Hoboken, NJ.

DiMasi, J.A., Grabowski, H.G., Hansen, R.W., 2016. Innovation in the pharmaceutical industry: new estimates of R&D costs. Journal of Health Economics 47 (5), 20–33.

Heller, M.A., 1998. The tragedy of the anticommons: property in the transition from marx to markets. Harvard Law Review 111 (3), 621–688.

Heller, M.A., Eisenberg, R.S., May 01, 1998. Can patents deter innovation? The anticommons in biomedical research. Science 280 (5364), 698–701. http://science.sciencemag.org/content/280/5364/698.

Heller, M.A., 1999. The Tragedy of the Anticommons. The Wealth of the Commons. http://wealthofthecommons.org/essay/tragedy-anticommons. Essay adapted from Chapter 2 of The Gridlock Economy, 2010. http://www.gridlockeconomy.com.

MassBio, July 26, 2017. Evolution of an Academic Discovery: First Person Account of the Development of Velcade®. http://www.ustream.tv/recorded/106250511.

Sánchez-Serrano, I., 2006. Success in translational research: lessons from the development of bortezomib. Nature Reviews Drug Discovery 5, 107–114.

Sánchez-Serrano, I., 2011. The World's Health Care Crisis: From the Laboratory Bench to the Patient's Bedside. Elsevier.

Sánchez-Serrano, I., October 10, 2012. El Modelo Core para el desarrollo de medicamentos. TEDx, Panama City. https://www.youtube.com/watch?v=ZQaq5_MJYjk (with English subtitles).

Sánchez-Serrano, I., 2014. The core model: drug discovery and development via effective translational science and public–private collaboration. In: Chaguturu, R. (Ed.), Collaborative Innovation in Drug Discovery. Wiley (Chapter 35). https://onlinelibrary.wiley.com/doi/book/10.1002/9781118778166.

Sánchez-Serrano, I., Pfeifer, T., Chaguturu, R., 2018. Disruptive Approaches to Accelerate Drug Discovery and Development (Part 1). Drug Discovery World. Spring. https://www.ddw-online.com/business/p322101-disruptive-approaches-to-accelerate-drug-discovery-and-development-(part-1).html.

Singer, C., Underwood, A., 1962. A Short History of Medicine. Clarendon Press.

Solow, R.M., 1956. A contribution to the theory of economic growth. Quarterly Journal of Economics 70, 65–94.

Stokes, D.E., 1997. Pasteur's Quadrant. Brookings Institution Press, Washington, DC.

Vagelos, R., Galambos, L., 2004. Medicine, Science, and Merck. Cambridge University Press.

Weatherall, M., 1990. In Search of a Cure: A History of Pharmaceutical Discovery. Oxford University Press, New York.

Index

'*Note:* Page numbers followed by "f" indicate figures, "t" indicate tables and "b" indicate boxes.'

D

E

I

J

K

L

M

Printed in the United States
By Bookmasters